CAMBRIDGE LIBRARY COLLECTION

Books of enduring scholarly value

Physical Sciences

From ancient times, humans have tried to understand the workings of the world around them. The roots of modern physical science go back to the very earliest mechanical devices such as levers and rollers, the mixing of paints and dyes, and the importance of the heavenly bodies in early religious observance and navigation. The physical sciences as we know them today began to emerge as independent academic subjects during the early modern period, in the work of Newton and other 'natural philosophers', and numerous sub-disciplines developed during the centuries that followed. This part of the Cambridge Library Collection is devoted to landmark publications in this area which will be of interest to historians of science concerned with individual scientists, particular discoveries, and advances in scientific method, or with the establishment and development of scientific institutions around the world.

The Midnight Sky

When Edwin Dunkin (1821–98) published this book in 1869, it was received with widespread acclaim by both professional astronomers and the reading public. Dunkin, a distinguished astronomer who published widely in academic journals and later served in the prestigious roles of Deputy Astronomer Royal (1881–4) and President of the Royal Astronomical Society (1880), is still best known for this work of popular astronomy that has functioned as an indispensable tool for generations of amateurs. Chapter 1 derives from Dunkin's famous 'The Midnight Sky at London' articles, previously published in *Leisure Hour*, which describe the London midnight sky during each month of the year. Other chapters cover the Southern Hemisphere, the constellations, the properties of fixed stars, the solar system, and meteors and shooting stars. The volume is well illustrated with star maps and engravings. It is a classic work of popular nineteenth-century astronomical writing.

Cambridge University Press has long been a pioneer in the reissuing of out-of-print titles from its own backlist, producing digital reprints of books that are still sought after by scholars and students but could not be reprinted economically using traditional technology. The Cambridge Library Collection extends this activity to a wider range of books which are still of importance to researchers and professionals, either for the source material they contain, or as landmarks in the history of their academic discipline.

Drawing from the world-renowned collections in the Cambridge University Library, and guided by the advice of experts in each subject area, Cambridge University Press is using state-of-the-art scanning machines in its own Printing House to capture the content of each book selected for inclusion. The files are processed to give a consistently clear, crisp image, and the books finished to the high quality standard for which the Press is recognised around the world. The latest print-on-demand technology ensures that the books will remain available indefinitely, and that orders for single or multiple copies can quickly be supplied.

The Cambridge Library Collection will bring back to life books of enduring scholarly value (including out-of-copyright works originally issued by other publishers) across a wide range of disciplines in the humanities and social sciences and in science and technology.

The Midnight Sky

Familiar Notes on the Stars and Planets

Edwin Dunkin

CAMBRIDGE
UNIVERSITY PRESS

CAMBRIDGE UNIVERSITY PRESS

Cambridge, New York, Melbourne, Madrid, Cape Town, Singapore,
São Paolo, Delhi, Dubai, Tokyo

Published in the United States of America by Cambridge University Press, New York

www.cambridge.org
Information on this title: www.cambridge.org/9781108017459

© in this compilation Cambridge University Press 2010

This edition first published 1869
This digitally printed version 2010

ISBN 978-1-108-01745-9 Paperback

THE

MIDNIGHT

SKY

THE

MIDNIGHT SKY:

FAMILIAR NOTES ON THE STARS AND PLANETS.

By EDWIN DUNKIN,

OF THE

ROYAL OBSERVATORY, GREENWICH, AND FELLOW OF THE ROYAL ASTRONOMICAL SOCIETY, LONDON.

With Thirty-two Star-Maps and numerous other Illustrations.

LONDON:

THE RELIGIOUS TRACT SOCIETY;

56, PATERNOSTER ROW; 65, ST. PAUL'S CHURCHYARD:

AND 164, PICCADILLY.

"The heavens declare the glory of God;
And the firmament showeth His handywork."

CONTENTS.

LIST OF ILLUSTRATIONS.

PREFACE.

—♦—

THE favour with which the articles and star-maps, entitled "The Midnight Sky at London," were received by the readers of the *Leisure Hour*, has encouraged the Committee to re-issue them, with some additional astronomical papers, in a separate and more convenient form. In the arrangement of this work for the press, the author has made a complete revision of every part, omitting many paragraphs which had reference only to the time of the serial publication, and adding much important matter which, in a permanent book, should have a place. He has not only re-written many portions of the text, but he has also inserted some remarks on the Solar System, including the Sun, the Earth, the Moon, Mercury, the Minor Planets, Uranus, Neptune, and Comets, none of which were described in the original papers. Considerable additions have also been made in other parts of the work, especially in the chapter on "General Notes on the Fixed Stars."

With regard to the star-maps, the author believes that, with the assistance of the corresponding index-maps, the student will experience but little difficulty in recognising the principal stars and constellations in the heavens. There may be a partial failure at first, but by persevering in a systematic comparison of the diagrams with the sky, any intelligent person will soon conquer whatever difficulty may arise. That many persons have succeeded in recognising the principal objects, by the use of the sky-views, has been proved by the numerous gratifying communications which the author has received from correspondents resident in all parts of Great Britain, during the publication of the articles in the pages of the *Leisure Hour*.

Astronomy has for ages been one of the most popular of the sciences. What greater delight can be experienced by those who take pleasure in the contemplation of natural objects, than by becoming personally acquainted with the relative positions of the planets and principal stars; with the general contour of the different constellations; and with many other illustrations of God's creative wisdom as displayed among the starry hosts above and around us? "Why did not somebody," says Carlyle, "teach me the constellations, too, and make me at home in the starry heavens, which are always overhead, and which I don't half know to this day?" Now, it has been the earnest desire of the author to be the "somebody" of Carlyle, and if his very familiar notes have enlightened the minds of his readers, even if only in a small degree, the numerous views of the sky and the illustrative descriptions will not have been prepared in vain.

E. D.

GREENWICH: *October* 1, 1869.

Greek letters being used to distinguish the principal stars, the following alphabet may be helpful to some readers :—

α	Alpha.	ι	Iota.	ρ	Rho.
β	Beta.	κ	Kappa.	σ	Sigma.
γ	Gamma.	λ	Lambda.	τ	Tau.
δ	Delta.	μ	Mu.	υ	Upsilon.
ε	Epsilon.	ν	Nu.	φ	Phi.
ζ	Zeta.	ξ	Xi.	χ	Chi.
η	Eta.	o	Omicron.	ψ	Psi.
θ	Theta.	π	Pi.	ω	Omega.

THE

MIDNIGHT SKY AT LONDON.

" There they stand,
Shining in order, like a living hymn
Written in light."

N. P. WILLIS.

INTRODUCTORY AND EXPLANATORY.

 FEW brief and elementary explanations will enable the reader to understand the method adopted in the construction of the following representations of the starry heavens as seen at midnight.

Although all the star-maps of the northern hemisphere are given for London exclusively, yet they are equally available for every place on the surface of the Earth where the latitude is 51° 30′ N. The selection of London is a purely arbitrary one; any other city in the same latitude would have answered our purpose as well. In all districts situated within a few degrees north or south of latitude 51° 30′ N., the star-views are also approximately correct, so that an ordinary observer, without a telescope, would probably fail to detect the difference. It is of no consequence, therefore, in what part of London, or its neighbourhood, the observer is located. It may be in the heart of the City, on the heights of Hampstead and Highgate, among the denizens of Kensington or Belgravia, or in the astronomical atmosphere of Greenwich. At all these places, the appearance of the heavens at midnight is, for all practical purposes, the same. And with slight and unimportant differences, which easily admit of correction, the same remark holds good of the whole of Great Britain. In the sky-views of London, the observer is supposed to be looking north, directly over St. Paul's Cathedral, and looking south over Greenwich Hospital, with the Royal Observatory in the background.

In the representations of the southern heavens, different places have been selected on the parallel of 34° S. latitude, including views in the Cape of Good Hope, Australia, and South America. In one or two instances, the position of the stars is as seen at sea from on board ship.

With respect to places in the same parallel of latitude, but separated by a considerable difference of longitude, there is one important circumstance which the reader must bear in mind. As all the diagrams of the sky of the northern hemisphere represent the view of the heavens at the *local* midnight of London, and those of the southern hemisphere the *local* midnight at the respective localities, so, if we wish to

3

compare them with the sky at other places in the same latitude, say in North America, Northern Germany, or even in Asia, it must be at the *local* midnight of each station. Or, if we desire to make a similar comparison with the southern sky for different places, it must also be at the *local* midnight of each locality. Let us take as an example the February diagrams of the southern sky, which correspond strictly with the heavens on February 15th at midnight, Cape mean time. Here the landscapes are views of two stations in Cape Colony, but an observer in Australia would have seen exactly the same stars above the horizon at his *local* midnight, which occurred eight or nine hours before. In this interval of time, the Earth has turned on its axis a distance corresponding exactly to the difference of longitude between the two countries. Wherever the reader of these pages is situated, therefore, no confusion need arise in his mind on this ground, for he has only to consider that the diagrams represent the heavens at *his own* midnight, whatever longitude his station may be, on the parallels of 51° 30′ north latitude, and 34° south latitude respectively. In Great Britain, Greenwich mean time, determined daily at the Royal Observatory by the observations of the transits of standard stars over the meridian, or railway time, as it is more commonly called, is now so universally kept in all parts of the country, owing to the extension of the railway system into its remotest corners, that the great majority of people know little or nothing of their own local time, as indicated by the culmination of the Sun. The adoption of a universal clock-time has been acknowledged on all sides to be a valuable contribution to the order and punctuality of the nation, without any corresponding drawback being produced on domestic interests by the difference of longitude between the extreme eastern and western coasts, which does not, at the greatest, exceed half-an-hour. If our diagrams were consulted in the Land's End district of Cornwall, the time selected for the comparison should be twenty-three minutes past twelve, railway time.

For ordinary purposes, it would not be convenient to adopt a universal time for different countries ; for if Greenwich mean time were kept all over the globe, although some advantages might occasionally be derived by the mariner for nautical purposes, the ordinary civil days would begin and end so contrary to all our notions of common sense, and so antagonistic to the natural day of the place, that endless confusion in the daily habits of the people would be the consequence. Every civilised country, therefore, has generally its own local or national time, referred in most cases to that of the metropolis, or of the city or town where the principal observatory of the nation is situated. In large and thinly-populated countries, where, in the absence of railways, intercommunication is difficult, the local time of the chief town of each province or district is frequently adopted. In Great Britain, as stated above, the standard is Greenwich mean time, communicated daily from the Royal Observatory to all parts of England by means of the telegraph wires in connection with that establishment. Irish time—the same as Dublin time—is used throughout our sister island, even to the westernmost coasts of Kerry. The whole of France keeps Paris time; Germany keeps Berlin or Vienna time; northern Russia, St. Petersburg or Pulkowa time; the United States of America, Washington, New York, or Boston time, and so on.

In like manner, in the southern hemisphere, at the Cape of Good Hope there is Cape time; in Australia there is probably Sydney, Melbourne, or Adelaide time, as the case may be. It can thus be easily perceived that, after the diagrams of the southern hemisphere have served their purpose as guides to the star-watcher at an Australian midnight, they will, eight or nine hours afterwards, be equally available for a Cape of Good Hope midnight, the configurations of the stars being seen in both places under exactly similar circumstances.

A few brief general explanations of the maps are requisite. In comparing the diagrams with the sky, the observer is supposed to be looking either *due* north or south along the exact meridian of his station. This position is absolutely necessary; for the apparent aspect of the constellations is different if the face be turned aside from the true plane of the meridian towards the south-west or south-east, north-west or north-east. One of the first objects to consider, therefore, is to be certain that the direction of the imaginary meridian line is approximately known. This may be done by several methods. In the day-time it can be determined by noticing the exact time when the Sun is on the meridian. As this takes place generally some minutes before or after ordinary local noon, reference must be made to the calendar of any almanack in which this difference is usually inserted. The moment when the Sun is in this position, it is at the highest point in the heavens, or on the meridian. At night, the same can be done by observing the Moon on the meridian, the time of which is also given in most almanacks. But the most useful method of all, because it may be adopted at any instant during the night hours, is by first finding the North Pole, indicated approximately by the position of Polaris, or the Pole star. This well-known fixed star can be easily detected by using the first two stars in the principal group of the constellation Ursa Major, as pointers, and is so near the celestial pole that, for all practical purposes of this kind, it may be assumed to be a stationary object. When the observer has identified Polaris, all he has to do is to draw an imaginary line in his mind from the pole to the zenith, and thence to the south horizon, and he has determined the meridian line* of his station sufficiently well for the comparison of the diagrams with the heavens.

Each of the star-views represents the sky included between the zenith and the north or south horizon, according as the observer is looking north or south. Practically this is more of the heavens than the eye can grasp at any one time, for the stars situated in and near the zenith cannot be seen without a special effort of the observer. As the celestial vault is spherical, it is very difficult to represent its appearance efficiently on a flat surface without considerable distortion of the constellations in some directions. There

---

* The word meridian is derived from the Latin *meridies*, mid-day, because when the Sun is in that position there is noon at all places of the same longitude as that where it is situated in the zenith. The term meridian may, however, be more exactly defined as a great circle of the celestial sphere passing through the zenith and the poles.

are several methods of constructing celestial maps, but there is not one in which the constellations are free from distortion, either more or less. In the projection adopted in the formation of this series of diagrams, this distortion takes place near the horizon, but not in the central part of the map. Partly for this reason, but chiefly because the scale of the diagrams would become far too small for the easy identification of the different objects, the sky low down in the east and west has been omitted. When, however, stars of superior magnitude are visible in the east and west out of the range of the maps, their positions are generally pointed out in the monthly descriptions, so that it will not be a difficult task for our youngest reader to identify them. The positions of the Moon and planets cannot of necessity be inserted, on account of the varying and peculiar character of their movements.

If the upper parts of the two monthly diagrams be joined together, the heavens from the south horizon to the north horizon will be represented. The reader is therefore requested to bear in mind that the upper boundary lines are due east and west, passing through the zenith, or the point in the heavens directly over-head, and that the meridian-line divides the diagrams into two equal portions. The central part of each is therefore on the meridian, about forty-five degrees from the zenith. Every star down to the fifth magnitude is inserted in the diagrams looking south; the northern diagrams contain all stars down to the fourth magnitude, and in some of the constellations to the fifth. The addition of many more small stars would only tend to produce confusion, without being of any practical benefit to the reader. The interest of most non-telescopic star-gazers is principally centred in those stars only which are prominent objects and easily to be distinguished.

The reader will find the index-maps a considerable assistance to him for the proper elucidation of the large diagrams, independently of the descriptive explanatory notes. In the preparation of the index-maps, care has been taken to include all the large stars of the first and second magnitudes, and many of the third. By the aid of these supplementary maps, every important star can, with very little trouble, be directly identified, not only in the large maps, but also in the heavens itself. The descriptive notes for each month must be considered complete in themselves, without reference to those in any other month, unless specially mentioned. For this reason, we have found it occasionally necessary to include a few remarks which involve a repetition of what has been given in a preceding month.

The apparent changing positions of the stars, with respect to the zenith and horizon in different seasons of the year, cannot fail to attract the notice of the most superficial observer. In the descriptive notes for March, the cause of these seasonal changes is briefly explained. The change is going on regularly to the extent of about four minutes of time daily, amounting in a month to two hours. This monthly alteration in the position of the stars at midnight is very appreciable in the successive diagrams. As the apparent positions of the stars thus change relatively to the meridian at midnight, or indeed at any fixed hour, in the different months of the year, we have selected a day in the middle of each month for their delineation. But although the star-maps in January represent the appearance of the heavens at midnight on the 15th of that month, yet the same maps are equally applicable at other night-hours in subsequent months of the year. The same remark may be repeated for each diagram. For the convenience of those who may use them at any other hour than midnight, we have prepared the following tabular statement, which gives, at one view, the hour and month when each diagram of the series is available for comparison with the sky :—

DIAGRAMS FOR MIDNIGHT IN	CORRESPONDING MONTH AND HOUR WHEN THE SAME DIAGRAMS ARE AVAILABLE.					
	6 P.M.	8 P.M.	10 P.M.	2 A.M.	4 A.M.	6 A.M.
January	...	March	February	December	November	October
February	...	April	March	January	December	November
March...	...	May	April	February	January	December
April	...	June	May	March	February	January
May	...	July	June	April	March	February
June	September	August	July	May	April	March
July	October	September	August	June	May	April
August	November	October	September	July	June	...
September	December	November	October	August	July	...
October	January	December	November	September	August	...
November	February	January	December	October	September	...
December	March	February	January	November	October	...

JANUARY.

—◆—

HILE comparing this series of diagrams with the sky, the observer is supposed to be stationed in some position where he can command an uninterrupted view of the heavens from the north to the south horizon.   To do this properly, he must be out of doors.   Viewing the sky from a window of a dwelling-house is very deceptive, as about thirty degrees of the sky from the zenith is completely cut off, making the apparent altitude of the stars above the horizon much greater than the reality.   In our remarks, we therefore assume that the observer has selected a favourable position, and that he has furnished himself with a hand-lamp, to enable him to make his comparison of the diagrams directly with the sky.   The first thing for him to do is to find the direction of the meridian approximately.   In the introductory chapter it has been stated that, in the night-time, the best method of fixing this imaginary line, is to discover the position of the North Pole, indicated very nearly by the well-known isolated star of the second magnitude, Polaris. When found, a little time will be well spent in forming some acquaintance with the general appearance of that part of the sky, so that, in future observations, the eye may fall readily upon the spot without special reference to other stars.

The most popular, and by far the easiest, way of finding Polaris, is to glance at the universally-known seven stars in Ursa Major, a constellation which has probably attracted the attention of most persons.   There can be no difficulty in identifying this Ursa Major group, the general form of which is so well understood.   Now, at midnight in January, the two stars nearest to the zenith, or point over-head, have for ages been popularly designated the Pointers, because a straight line drawn from Merak, the more southerly, to Dubhe, the more northerly of the two, will lead, if continued, very nearly to Polaris. These two leading stars therefore clearly indicate the position of the Pole of the heavens and the Pole star, which is situated in a part of the sky free from other stars of a similar magnitude.   Probably there will be little chance of mistaking any other object for Polaris after reading the following lines :—

" Where yonder radiant hosts adorn
    The northern evening sky,
  Seven stars, a splendid glorious train,
    First fix the wand'ring eye.

To deck great Ursa's shaggy form,
    Those brilliant orbs combine ;
And where the *first* and *second* point
    There see *Polaris* shine."

THE MIDNIGHT SKY AT LONDON, LOOKING NORTH, JANUARY 15.

INDEX-MAP.

E

W

THE MIDNIGHT SKY AT LONDON, LOOKING SOUTH, JANUARY 15.

INDEX-MAP.

Having found Polaris, we will now endeavour to make a detailed comparison of the southern diagram with the sky south of the zenith, which it is hoped the reader will be able to follow easily. First, however, let us find our meridian by means of Polaris. While we have, in fact, been noticing this star, we have been unconsciously looking along the plane of the meridian, north of the zenith. As the meridian line is a great circle passing through the zenith and the poles, it is evident that if we turn our back towards the Pole star, and carry our eyes downwards from the zenith to the south horizon, we shall describe an arc which will be the meridian of the observer's station. On January 15, at midnight, the accuracy of the determination can be at once verified, for on that day and hour the three bright stars, Castor, Pollux, and Procyon, are very near the south meridian.

The direction of the meridian being duly settled, we commence the comparison of the southern diagram with the heavens, by looking directly overhead, which corresponds exactly with the centre of the boundary line from east to west in the upper part of the diagram. The zenith is now occupied by the constellation Lynx, which contains very few stars above the fifth magnitude, consequently that part of the heavens looks comparatively bare. Passing slowly downwards on the meridian, we soon come to the two bright objects Castor and Pollux, the latter being the more southerly. Lower down, about forty-five degrees from the zenith and horizon, Procyon, in the loins of Canis Minor, shines out as a very prominent star. One a little north-west of Procyon is Beta Canis Minoris. The following constellations are at this time on the meridian, commencing at the zenith—Lynx, Gemini, Canis Minor, Monoceros, and Argo Navis, the last being near the horizon.

West of the meridian, west and south-west of Castor and Pollux, the most interesting portion of the sky of the northern hemisphere is spread out before us. First, we have Taurus, with its principal star, Aldebaran, surrounded by numerous smaller gems, commonly called the Hyades. North of these are the Pleiades, so easily recognisable to the naked eye by their nebulous appearance. Then there is the beautiful constellation of Orion, second to none in either hemisphere, but equally visible in both; while nearer the south horizon and the meridian, the splendid dog-star Sirius, with its very brilliant attendants, are all included in Canis Major. Looking due west, Capella, twenty-five degrees from the zenith, is very conspicuous, followed by Beta Aurigæ. These two stars being circumpolar, never set below the horizon of London; they are consequently visible, like the seven stars in Ursa Major, on every night throughout the year. Capella being slightly north of west at midnight in January, is inserted in the map of the northern sky. The numerous stars between Castor and Pollux and Aldebaran belong partly to Taurus, and partly to Gemini. A tolerably bright object near the western edge of the Milky Way is Beta Tauri. Of the Orion group, the two brightest stars are Betelgeuse and Rigel, the first being that nearest the meridian, and the latter that nearest the horizon. The three in the belt can be found by any one at a glance. Below Orion, near the horizon, are Lepus and

Eridanus. Cetus is very low down in the west. The Milky Way adorns the western sky at this time, adding considerably to the brilliancy of the surrounding objects.

The sky east of the meridian can bear no comparison with that which has just passed under our notice. The principal group is that of Leo, which contains Regulus, Denebola, and several other large stars. These can all be identified in the upper part of the left-hand side of the diagram. The index-map gives the name of each. Some distance below Regulus, the chief star in Hydra is easily seen, being rather conspicuous on account of the absence of any object of equal magnitude within a considerable distance around. The principal constellations above the horizon in this quarter of the sky at this time, are Cancer, Leo, Hydra, Sextans, Leo Minor, Crater, and portions of Lynx, Ursa Major, Coma Berenices, Virgo, and Argo Navis.

The observer is now requested to turn round so as to face Polaris. He will then perceive that between that star and the zenith, there is not a single object worthy of special notice. This district is occupied by the unimportant constellations Lynx and Cameleopardalis, and by an outlying portion of Ursa Major. We must not confine this last constellation to the limited area occupied by the seven principal stars, for it extends south and west of them for some distance. This group is now approaching the meridian, over which the two advanced stars, being nearly of the same right ascension, pass almost simultaneously. When Alkaid, in the tip of the Bear's tail, is in this position, it is about three degrees south of the zenith. Near the horizon in the east, out of the range of the map, Arcturus has just risen. Several other stars in Boötes are also visible. Looking very low down, in the direction of Ursa Major, the Northern Crown has also just risen. The horizon in this quarter of the sky is principally occupied, from east to west, by Boötes, Corona Borealis, Hercules, Lyra, and Cygnus. Alpha Cygni, the chief star in Cygnus, is visible in the diagram over the western towers of St. Paul's. East of Polaris, two average stars mark the extreme limit of Ursa Minor, the general contour of which can be clearly made out on very fine nights, the other extremity of the constellation being near Polaris, which is situated in the Lesser Bear's tail. Between Ursa Minor and Hercules or Lyra, some of the chief stars in Draco may be identified, especially Beta and Gamma Draconis. Reference should be made to the index-map for their names. The brilliant star, Alpha Lyræ, frequently called Vega, may be perceived at midnight, when the horizon is free from haze, about one degree high.

West of Polaris, in the Milky Way, the group in Cassiopeia can be easily recognised. Another group, also in the Milky Way, above Cassiopeia belong to Perseus, the variable star Algol being in a small offshoot of that luminous stratum. Cepheus is below Polaris, between Cassiopeia and Draco. A portion of Auriga occupies the upper part of the map west of the zenith; its position is marked by the first-class star Capella. The horizon from north to west contains portions of Cygnus, Andromeda, and Pisces. Alpha and Beta Arietis, in the constellation Aries, are low down in the west, out of the range of the diagram. The Milky Way in the Northern half of the sky passes through Perseus, Cassiopeia, Cepheus, and Cygnus, where it sets below the north horizon.

14

The preceding remarks, referring to the time of midnight on January 15th, are also strictly applicable to four minutes before midnight on January 16th ; eight minutes before midnight on January 17th, and so on. As stated in the introductory chapter, the appearance of the heavens thus changes in relation to the meridian at any stated time, about four minutes each day, a whole year passing away before the same star can culminate at the same hour of solar time. For example, the stars which pass the meridian at midnight on January 15th, 1870, will not pass again at that hour till about January 15th, 1871. The cause of these changes is more fully explained in the March notes. Thus the diagrams for midnight of each month serve for other hours, as given in the tabular statement at the end of the introductory chapter.

The principal constellations of the north sky are all circumpolar, and consequently, in the latitude of Great Britain, their chief stars never set below the horizon. They pass the meridian twice during the twenty-four hours,—once above and once below the Pole. They can, therefore, be seen at midnight, or, indeed, at any hour throughout the year when the sky is clear. At any fixed time they may be noticed at one season of the year in the east, at another season in the west, at other times north or south of Polaris. An examination of the complete series of diagrams of the north sky shows at sight these seasonal changes of position for the hour of midnight. But he must be a very careless observer who has not also noticed similar daily changes of position arising from the Earth's rotation on its axis once in twenty-four hours. As an example of the daily apparent revolution of the circumpolar stars round the Pole, let us note the position of the Great Bear at six o'clock in the evening of any day in January ; at this time it will be found adorning the heavens towards the north-east horizon ; at midnight it will be as depicted in the diagram of the north sky ; at six o'clock in the morning it will be in a corresponding position on the opposite side of the meridian ; and at noon it will be towards the north-west horizon, when all its bright stars can be easily observed by the aid of ordinary telescopes. In observatories there is now no difficulty experienced in making astronomical observations of the principal objects in daylight ; on the contrary, most of the stars included in the diagram of the northern sky can be seen at all hours of the day when passing the meridian. In the days of Flamsteed, the first Astronomer Royal, ordinary telescopes were not sufficiently powerful for this purpose ; but in order to observe daily, if possible, a celebrated star, Gamma Draconis, which happens to pass over the zenith of Greenwich, he had a deep well excavated, in which he erected a long tube or telescope. In the archives of the Royal Observatory a contemporary drawing of this well is preserved, showing an observer, probably the celebrated Flamsteed himself, at the bottom, in the act of making an observation of this standard Greenwich star. At the present day Gamma Draconis is viewed by means of a beautiful and precise instrument, designed by the present Astronomer Royal, called " the reflex zenith-tube," placed in a small apartment level with the usual observing-rooms ; the observer has therefore no need to descend below the surface of the ground to see this star, whether it be in the day or night.

FEBRUARY.

———◆———

N comparing the diagram of the sky south of the zenith for midnight in the middle of February with that for the middle of January, it will at once be perceived that considerable changes in the position of the constellations, with relation to the meridian, have taken place, all the stars having, without exception, travelled towards the west. Beginning at the west, or right-hand side of the diagram, we find that the constellation Orion has, for the most part, passed out of its limits, though it is still visible in the heavens very near the horizon a little south of west. Throughout the evening hours this splendid constellation is the most conspicuous object in the heavens, passing the meridian, or highest point, about seven o'clock, and then gradually descending to the horizon, where it sets about one A.M. Following Orion, the brilliant star Sirius is disappearing from view at midnight. Castor, Pollux, and Procyon, all of which were on the meridian in January, have now considerably advanced towards the west, while the large star Regulus, in Leo, is near the meridian. The stars Arcturus in Boötes, and Spica in Virgo, are just entering as brilliant objects, one ruddy, the other white, Arcturus being distinguishable due east, and Spica in the south-east. The small constellation Corvus, low down in the south-east, is soon identified by its four moderately bright stars. No object worthy of notice is near the zenith, excepting a few stars of the third magnitude in Ursa Major, a part of which at this hour is directly overhead. The principal stars of this constellation are, however, north of the zenith, including all the seven stars commonly called Charles's Wain, or the Plough.

We are now looking directly south, towards the meridian, the exact position of which we have determined by the rules given in January. We may reasonably assume, therefore, that there will be no more difficulty in ascertaining the true line of north and south. If we look upwards near the zenith, a few stars in Ursa Major will strike the eye. A close pair of stars a little west of the zenith are Iota and Kappa Ursæ Majoris in the Great Bear's right fore-foot. A similar pair, a little lower, and nearly on the meridian, are two rather brighter stars, named Lambda and Mu Ursæ Majoris, situated in the Bear's right hind-foot. These two pairs of stars are very clearly distinguished from the surrounding small objects, not only when in their present position, but also when in the north-western and north-eastern quarters of the sky. Passing the eye down the meridian we come

16

THE MIDNIGHT SKY AT LONDON, LOOKING NORTH, FEBRUARY 15.

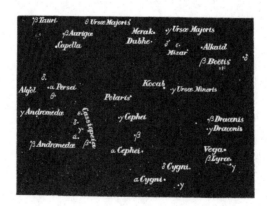

INDEX-MAP

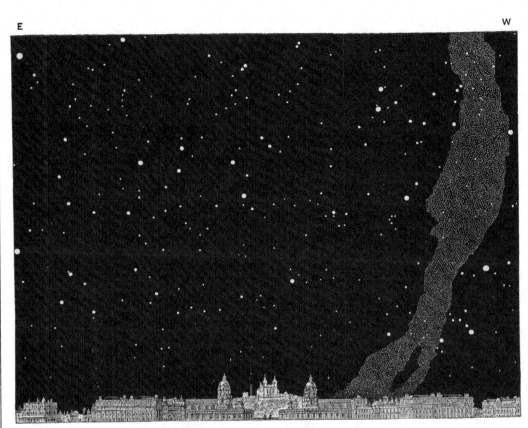

E

W

THE MIDNIGHT SKY AT LONDON, LOOKING SOUTH, FEBRUARY 15.

INDEX-MAP.

to an interesting group in Leo, popularly named the Sickle, from its resemblance to the form of that agricultural instrument. This group consists of several stars, including Regulus, of the first magnitude, Gamma Leonis, of the second magnitude, and Epsilon, Eta, and Mu Leonis, of the third magnitude. Almost due east of the Sickle the next two moderately sized stars are Delta and Theta Leonis. East of these is Denebola, marking the position of the tail of Leo. Denebola, Arcturus, and Spica, form very nearly an equilateral triangle, which is an easy means for the identification of each. The following not very poetical lines refer to these stars :—

> "From Deneb, in the Lion's tail,
>   To Spica draw a line,
> Then will these two with Arcturus
>   A bright triangle shine."

Between Leo and the nearest close pair of stars in Ursa Major, the small constellation Leo Minor is situated. It contains no star greater than the fourth magnitude. Between Denebola and Spica, several stars of the third magnitude belong to Virgo. Hydra extends from near Canis Minor, west of the meridian, to below Spica, near the south-east horizon. Alphard, its principal star, and the only one at all conspicuous, is slightly west of the meridian, south-west of Regulus. Sextans, between Leo and Hydra, is a small constellation containing only stars of inferior magnitude. Corvus and Crater are between Virgo and Hydra. The comparative dearth of large stars in the neighbourhood of Alpha Hydræ presents a great contrast to the brilliancy of the more northerly constellations.

We have stated above that many of the most attractive groups of stars are now passing away in the west and south-west. On this side of the meridian, Castor and Pollux, and several bright stars included in Gemini, are the principal objects between the zenith and the brilliant groups of Orion and Canis Major, now near the W.S.W. and S.W. horizon. The Milky Way may be seen passing between Procyon and Sirius, through the unimportant constellation Monoceros.

In the diagram, looking north, all the circumpolar stars included in that for January 15th can again be easily recognised, although in relation to the Pole and meridian they exhibit a sensible alteration. Ursa Major is still on the eastern side of the meridian, to which it is gradually approaching. To view Ursa Major at this time, we must look nearly over-head ; the most convenient way is to lie in an inclined position, with the face directed towards the zenith. The other circumpolar constellations, Perseus, Cassiopeia, Cepheus, Draco, and the most distant of the stars from the Pole in Ursa Minor, have advanced, in the same proportion as Ursa Major, in the direction from west to north and east, while Andromeda, Cygnus, Lyra, Hercules, and others, being only partially circumpolar at London, have moved in the same direction at greater distances from the Pole. Since January 15th, a considerable portion of Auriga has passed from the southern half of the sky to the northern.

Looking towards the west, or rather north of west, the two principal stars in Auriga, Capella and Beta Aurigæ, are the first objects which will attract attention. These two stars, being never absent from our view on starlight nights, are known almost as well as the Ursa Major group. They are now about half-way between the zenith and horizon, Capella being the larger and farther from the zenith. Lower still, but more towards the north-west, the group of stars in Perseus can be pointed out without difficulty, especially if the index-map be referred to at the same time. Of the Perseus group, Algol is at the western edge of the Milky Way, Alpha Persei being in the centre of that luminosity. The Cassiopeia group is in the N.N.W., being always in opposition to Ursa Major, in relation to the Pole. West of Cassiopeia, near the horizon, a part of Andromeda can be distinguished. Cepheus is on the meridian below Polaris. On clear nights, the stars forming Ursa Minor can be easily traced, by their similarity of arrangement to those of Ursa Major; the two terminating stars, Beta and Gamma Ursæ Minoris, are, however, much brighter than those between them and Polaris.

Below Ursa Minor, in the direction of the horizon in the north-east, several bright stars in Draco are distinctly visible. Beta and Gamma Draconis are above Vega. It may be remarked here, in passing, that as Polaris is the nearest conspicuous star to the pole of the equator, so Zeta Draconis, of the third magnitude, is the nearest to the pole of the ecliptic. Near the horizon in the N.N.E., Alpha Cygni, sometimes called Deneb, is visible. Proceeding onwards near the horizon in an easterly direction, we pass over Lyra, Hercules, Corona Borealis, and Boötes, the last constellation occupying the eastern horizon. Between Boötes and the zenith we come across once more the most attractive constellation of the northern sky, Ursa Major. In the early hours of the evening this group will be situated in the north-east. Using it as a point of reference, the positions of all the circumpolar constellations can be found with little trouble.

MARCH.

THE monthly variations in the positions of the stars, in relation to the meridian, zenith, or horizon, are so evident to the sight of the most indifferent observer, that it may be useful in this place to give a brief explanation of these apparent changes in the aspect of the stars, which are produced by corresponding changes in the relative positions of the Earth and Sun in the ecliptic. Before describing, therefore, the appearance of the midnight sky of March, we will, in a few familiar words, endeavour to state the cause of these seasonal changes in the appearance of the heavens at any fixed hour. We shall first suppose that our readers clearly understand that the Earth revolves on its axis in about twenty-four hours, and that a civil or solar day is measured by the time elapsed between two successive transits of the Sun over the meridian of any place. We shall also assume that the cause of the hourly changes in the positions of the stars with respect to the meridian is generally known to be simply the result produced by the diurnal motion of the Earth on its axis. Were the time of rotation exactly twenty-four hours of solar time, there would be no seasonal changes; for then any one star would always culminate at the same hour throughout the year. But the interval between two successive returns of any fixed star to the same meridian is smaller by several minutes than that given by consecutive transits of the Sun. Hence we have two days of different length, one the sidereal and the other the solar day. Now, we wish to explain briefly the cause of this difference. In the interval between two successive transits of the Sun, the Earth has traversed through about one degree of its orbit from east to west; this makes a corresponding apparent change in the position of the Sun in the heavens from west to east. The effect of this is, that the Earth has to make rather more than one revolution on its axis daily before the centre of the Sun can again coincide with the meridian; consequently, as we always reckon the length of day, according to our habit, by the daily passage of the Sun over the meridian, a solar twenty-four hours must be so much longer than a sidereal twenty-four hours by just the time occupied by the Earth's rotation in the interval of time equal to the change in the position of the Sun on the ecliptic since the preceding noon. But the apparent motion of the Sun is not always uniform, the amount of its daily increasing right ascension depending upon the portion of the ecliptic in which it is situated. For the convenience of daily use, it being impossible to construct a time-piece to follow the course of the Sun entirely, the mean, or average, motion of the Sun is adopted instead of the true motion, so that the interval between two consecutive transits of the mean Sun shall equal exactly twenty-four hours of clock-time all the year round. The difference of time between the moment when the Sun is really on the meridian, and when the fictitious or mean Sun would be in that position, is called "the equation of time," the daily amount of which is generally inserted in ordinary almanacks as "Clock before or after Sun." A mean solar day, or the day which is indicated by all clocks and watches, is longer than a sidereal day by about $3^m 56^s$, or in other words, the length of a sidereal day is $23^h 56^m 4^s$ of ordinary clock-time. The interval, $23^h 56^m 4^s$, between two successive transits of a fixed star over the meridian of any place, marks therefore the time occupied by the simple diurnal rotation of the Earth, without any reference to the position or movements of the Sun. So far then as the fixed stars are concerned, the aspect of the heavens is always the same at any one sidereal hour in whatever part of its orbit the Earth may be situated. For it must not be forgotten that the distances of the fixed-stars are so great, that even a change in the position of the Earth in its orbit from one end of the ellipse to the opposite end, separated from each other as they are by about 184 millions of miles, makes no sensible difference in the relative positions of the stars in the sky, even when viewed with the most delicate instruments. A sidereal day being thus shorter than a mean solar day by nearly four minutes, it will be evident to every one that when any star passes the meridian at midnight on any day —say March 15th—it will on March 16th pass about four minutes before midnight, and on March 17th eight minutes before that hour, and so on. From this it will be seen that in course of time the star will pass the meridian at a much earlier hour, till it is lost in the daylight. After the lapse of one year exactly, it will again culminate about midnight on March 15th. What is done in the case of one star, is done for all, as their relative positions are always the same.

By a comparison of our diagrams for March with those for January and February, these apparent movements of all the stars from east to west can be readily distinguished. Taking Regulus as an example, it will be perceived that this bright star at midnight on January 15th is some distance east of the meridian; on February 15th it is nearly in the centre of the diagram, or near the meridian; while on March 15th it is considerably to the west. In like manner, all the stars in the south diagram have changed their positions in the same direction, some having disappeared in the west and south-west, while others have become visible in the east and south-east. If we wish to examine the heavens at an earlier hour than midnight—we will suppose eight o'clock in the evening—the observer has nothing more to do than to take the January diagrams, with the accompanying explanation, when he will be able to identify the principal stars in the same manner as at midnight in January. Again, if at ten o'clock in the evening, the February diagrams will be perfectly available for the purpose. For instance, Castor, Procyon, and Pollux are due south at midnight on January 15th; on February 15th they will be in a similar position at ten o'clock, and on March 15th at eight o'clock. For succeeding months, the reader is referred to the tabular statement at the end of the introductory chapter.

THE MIDNIGHT SKY AT LONDON, LOOKING NORTH, MARCH 15.

INDEX-MAP

E   W

THE MIDNIGHT SKY AT LONDON, LOOKING SOUTH, MARCH 15.

INDEX-MAP.

Returning to the midnight sky of March, and remembering that the centre of the upper portion of the diagrams is directly over the head of the observer, we will direct our attention, in the first place, to the sky south of the zenith. Several stars of the first magnitude will be at once noticed in different directions. Due west, rather more than half way towards the horizon, Castor and Pollux can be distinguished above any other object near them. It so happens, however, that if we draw an imaginary line from east to west through the zenith, Castor will be found to be in the northern division of the heavens, and Pollux in the southern. In our illustrations, therefore, Castor is inserted in the north diagram, while Pollux remains in the south. Proceeding towards the horizon in the W.S.W., Procyon, near the western limit of the diagram, is still shining brilliantly. Near the zenith, a few stars in Ursa Major can be easily identified. These, with the exception of Cor Caroli, are the brightest stars in a considerable region south of the zenith. The great constellation Ursa Major, if we except the well known seven stars in Charles's Wain, is not celebrated for the possession of many stars of large magnitude. Nearly one-half of Ursa Major is now south of the zenith. Of the two pairs of stars near the zenith, in the February diagram, looking south, one has passed into the northern half of the sky. That remaining in the south map consists of Lambda and Mu Ursæ Majoris. In the south-west, nearly midway between the zenith and horizon, Regulus is visible as the brightest star in that neighbourhood. Gamma Leonis and the other stars in the Sickle can still be favourably seen. Between Regulus and the meridian the space is occupied by the constellation Leo, Denebola being at the extremity of the Lion's tail, and exactly on the meridian. South-east of Leo is the small constellation known as Sextans, with no star greater than the fifth magnitude; and lower down in the same direction is the rambling, tortuous Hydra, which extends from near Procyon along the whole lower portion of the sky near the south horizon. Alpha Hydræ can be identified south-west of Regulus. The small constellations, Crater and Corvus, on the back of Hydra, are now near the meridian, Crater being a little to the west, and Corvus to the east. Corvus will be found with little difficulty, its four principal stars forming a trapezium at no great distance from the horizon, south-west of Virgo, east of the meridian.

On the eastern side of the meridian several important constellations and well-known stars are sure to attract our notice. Nearly over head, at a short distance from the zenith, the tolerably bright star Cor Caroli shines in a neighbourhood chiefly occupied with small objects. Passing the eye downwards, a little to the east of the meridian, we come to Coma Berenices (Berenice's Hair). It is not difficult to notice this constellation of small stars, situated exactly midway between Cor Caroli and Denebola. This group has a nebulous or woolly appearance to the naked eye. It contains numerous stars of small magnitude not inserted in the diagrams. A considerable portion of the south-east sky is occupied by Virgo, with Spica, a star of the first magnitude, and several of the third. One of these stars is near the meridian, immediately south of Denebola. This is known by the name of Beta Virginis. In the space between Denebola and Spica there are four other average stars in Virgo, which form, with Beta, very nearly two sides of a right-angled triangle. The

D   29 star next to Beta is Eta, then Gamma, a celebrated binary star, then Delta, the last being Epsilon Virginis. South-east of the corner star of this triangle the position of Spica is pointed out. East of Spica, near the south-east horizon, two bright stars in Libra have just risen. Nearly midway from the zenith to the horizon, in the E.S.E., Arcturus and several stars of the second and third magnitudes, in the constellation Boötes, are conspicuous objects; while about half-way between the zenith and horizon, looking due east, the interesting group in Corona Borealis can be distinguished by its principal member, Alphecca or Gemma. Below Boötes and Corona Borealis, Serpens and portions of Hercules and Ophiuchus are visible, the last constellation occupying the eastern horizon.

At midnight, in March, the following are the principal constellations above the horizon in the southern half of the sky: Cancer, Leo, Virgo, Libra, Canis Minor, Hydra, Sextans, Leo Minor, Crater, Corvus, Boötes, Serpens, Coma Berenices, and parts of Ursa Major, Corona Borealis, Hercules, Ophiuchus, and Gemini.

If we now refer to the diagram, looking north, we shall find that the appearance of the northern sky at midnight is sensibly different from that at the corresponding time in the preceding month, although, in consequence of the stars being for the most part circum-polar, many of the constellations may not appear to have changed their positions to any great extent. Over head, a part of Ursa Major is on the meridian, the pointers Dubhe and Merak being a little west of it, while the remaining stars in Charles's Wain are approach-ing the meridian. Rather more than one-half of this constellation is north of the zenith, including all the principal stars. Looking due west, Castor and Pollux are visible, Castor being north of the imaginary line separating the two halves of the sky. Pollux, as we have before mentioned, has not yet passed into the northern half, though it is on the point of doing so. The principal constellations between Castor and the north horizon, in this part of the sky, are Auriga, Perseus, and Cassiopeia. The two bright stars rather more than midway between the zenith and horizon, in the north-west, near the Milky Way, are Capella and Beta Aurigæ. Lower down, approaching the north, is Perseus, with its group of moderately bright stars, amongst which is the variable star Algol, in the head of Medusa; and near the north meridian, about twenty degrees above the horizon, the group of bright stars in Cassiopeia, below Polaris, can be easily recognised. On the eastern side of the meridian, below Ursa Major, the stars in Ursa Minor and Draco are clearly seen; below these is the constellation Cepheus. In the north-east, the brilliant Vega is the most conspicuous object, followed at some distance by Alpha Cygni, in the Milky Way. Hercules is in the E.N.E., about midway between the zenith and horizon. Polaris is of course sensibly in the same position as in February. A reference to the index-map, looking north, will point out the positions of the individual stars.

APRIL.

HE aspect of the midnight sky of April is somewhat different from that depicted in the diagrams for March, numerous stars of large magnitude having since last month made their appearance in the east and south-east, while some conspicuous objects in the west have disappeared below the horizon. The stars above the second magnitude, now visible south of the zenith, are Regulus in Leo, Spica in Virgo, Arcturus in Boötes, and Antares in Scorpio. Taken as a whole, the south sky at midnight still contains a considerable number of interesting objects, although no constellation has that striking appearance which Orion and some others present at an earlier or later period of the year. For example, we have still the bright stars in Leo and Virgo in the western part of the sky, while on the eastern side of the meridian those in Boötes, Serpens, and Scorpio are conspicuous objects at different altitudes. There are numerous other constellations, which we shall point out to the observer, who, with the assistance of our diagrams, can have no difficulty in identifying the principal stars and the constellations of which they are the chief members.

We must again suppose the observer standing with his back to the Pole star, with his eye directed along the plane of the meridian looking *due* south. At first, however, we would wish him to look directly over head, still with his back towards Polaris. He will then perceive that the last of the seven stars in Charles's Wain, in Ursa Major, is very nearly in the zenith; it is really about two degrees south of that point. This is the only month that this star, Eta Ursæ Majoris, or, as it is sometimes named, Alkaid and Benetnasch, is in the south half of the sky at midnight; in our diagrams for March and also for next month, it will be found with its six companions in the northern half. Passing the eye downwards, a bright star, with a reddish tinge, strikes our attention slightly east of the meridian; this is Arcturus. Considerably lower, a few degrees west of the meridian, the white star Spica shines brilliantly in a district in which it has no rival. The well-known nearly equilateral triangle formed by Denebola, Spica, and Arcturus, is best seen when Spica is near the meridian.

Let us now give a glance at the south-western division of the sky. Commencing at the zenith near Eta Ursæ Majoris, the first star of importance in this direction is Cor Caroli, the chief member of the small constellation of Canes Venatici (the greyhounds). It will be remarked that Cor Caroli is in an isolated position, no star of equal magnitude being near it for some distance on all sides. Then comes the group of small stars in Coma Berenices, below which, extending from the meridian to a considerable distance towards the west and south-west, Leo and Virgo occupy the major portion of the sky. The principal stars in Leo are all visible about midway between the zenith and horizon, or in the right hand upper corner of the diagram, looking south; they may be recognised in two ways, one by the irregular trapezium formed by the brightest objects, and the other by the rough form of a Sickle, alluded to in preceding months, given by the group of stars of which Regulus is the most southerly. The most favourable time for recognising the form of this celestial reaping-hook at this season of the year, is when it is situated near the meridian in the earlier hours of the evening, or about 8 P.M. in April, at which time it can be seen to great advantage and with considerable clearness. This group, near which is the radiant point of the November stream of meteors, consists of about six stars of average magnitude, one being between the first and second, one of the second, two of the third, and two of the fourth magnitude. Between Leo and the meridian, most of the chief stars in Virgo are situated, but, excepting Spica, none of them is greater than the third magnitude. One of these, Gamma Virginis, is, however, a most interesting binary star, perhaps the most carefully observed of any in the heavens. Some account of the relative movements of its components will be given in the descriptive notes of the constellation Virgo. The south-west portion of the sky towards the horizon is occupied chiefly with the small constellations, Sextans, Corvus, and Crater, while Hydra is spread over the horizon from west to nearly south.

East of the meridian, and at no great distance from it, the constellation Boötes, containing Arcturus and several stars of the second and third magnitudes, is the most attractive group in that portion of the heavens. Below Boötes, about half-way between the zenith and horizon in the south-east, is Serpens, and lower still, near the horizon, Scorpio will be recognised by several bright stars, of which the principal is Antares, of the first magnitude, and Beta Scorpii, of the second. This constellation is very prolific in stars of the fourth and fifth magnitudes, in addition to several of the third. Joining Boötes on the east, the small, but important, constellation Corona Borealis, can be noticed in the form of a semi-circular group, the central gem in the crown being its chief star Alphecca. Below Serpens, and between Scorpio and the meridian, Libra covers a considerable space. Its two principal stars are Alpha and Beta Libræ, both of the second magnitude. The horizon from east to south divides the constellations Aquila, Ophiuchus, Scorpio, Lupus, and Centaurus. The last two are rich in stars, but scarcely one of importance rises above the horizon of London. The south-west horizon is almost exclusively occupied by the contorted figure of Hydra, which joins Centaurus in the S.S.W., and Canis Minor very low down in the west.

The principal constellations visible at midnight in the northern sky, in addition to those which are wholly circumpolar, are Lyra, Cygnus, Lynx, Vulpecula, and parts of Cancer, Gemini, Auriga, Perseus, Andromeda, and Aquila. At this time, six of the seven

W

E

THE MIDNIGHT SKY AT LONDON, LOOKING NORTH, APRIL 15.

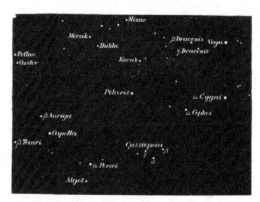

INDEX-MAP.

THE MIDNIGHT SKY AT LONDON, LOOKING SOUTH, APRIL 15.

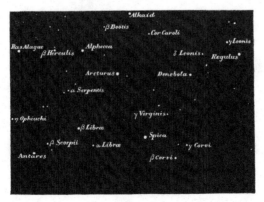

INDEX-MAP.

principal stars in Ursa Major are near the zenith, a little to the west of the meridian, the seventh in the tip of the tail being about two degrees south of the zenith. Below Ursa Major, towards the north-west, are the constellations Lynx and Cameleopardalis, in both of which there is scarcely any star greater than the fifth magnitude; consequently this portion of the heavens looks comparatively bare. A considerable number of small stars are, however, contained in these two constellations, though not inserted in the diagram. In the W.N.W., between Lynx and the horizon, Castor and Pollux can still be distinguished, and towards the north-west, near the Milky Way, Beta Aurigæ and Capella are very conspicuous. Near the north horizon, several bright stars in Perseus can be seen slightly west of the meridian, while about the same distance east, but nearer the pole, Cassiopeia is visible. Proceeding eastward, the stars of the first magnitude, Alpha Cygni and Vega, shine above all others near them; and almost due east, and very near to the horizon, the chief stars in Aquila have just risen. The strictly circumpolar stars, in Draco and Cepheus, can be generally recognised east of the meridian between Polaris, Alpha Cygni, and Vega. The stars in Ursa Minor are now nearly all between the zenith and the pole, Kocab, or Beta Ursæ Minoris, being near the meridian.

MAY.

———◆———

I F we refer to our diagram representing the sky south of the zenith for midnight in the middle of May, we shall discover that it is enriched by several of the most celebrated stars of the first magnitude, visible above the horizon of London, including Vega, Arcturus, Altair, Antares, and Spica. Many new constellations have appeared in the east and south-east since last month, while the general advance of all the stars from east to west is clearly marked by the altered positions of the brightest objects.

Taking, then, first, the diagram looking south, the attention of the observer is directed to the zenith, which is now occupied by the north-eastern corner of the constellation Hercules. It is the first time this year that Hercules is principally contained in the southern half of the sky at midnight. There are one or two stars of the second magnitude, and several of the third, in this constellation, some of which are near the zenith, east of the meridian line. Hercules extends over a large portion of this region of the heavens. West and south-west of Hercules, and on the meridian, the semicircular group of stars forming the Northern Crown is very conspicuous, its brightest jewel being Alphecca, or Gemma, in the centre. Directly below Corona Borealis is Serpens, with several stars of the second and third magnitudes; and, lower still, very near the horizon, is Scorpio, with its principal star Antares, and several others tolerably bright, including the double star Beta Scorpii. Looking due east, the brilliant Vega attracts our notice, about thirty degrees from the zenith. This star has passed since April 15th from the northern half of the sky to the southern. Near Vega there are several stars in Lyra of the third magnitude. Below Lyra is Beta Cygni, or Albiero, a double star celebrated for the number of observations made on the colours of its two components. Between Albiero and the eastern horizon are several small constellations, the principal being Vulpecula, Sagitta, Delphinus, and Equuleus, the horizon itself being occupied by the sign Aquarius. In the E.S.E. the constellation Aquila can be distinguished by its group of three stars, the central one being Altair, of the first magnitude. In the diagram, Altair is inserted near the eastern limit; it is, however, about thirty degrees above the horizon. The horizon in the south-east and south is occupied, beginning from Aquarius in the east, by Capricornus, Sagittarius, Scorpio, and Lupus. If we now turn to the west and south-west sky, we shall

THE MIDNIGHT SKY AT LONDON, LOOKING NORTH, MAY 15.

INDEX-MAP.

E   W

THE MIDNIGHT SKY AT LONDON, LOOKING SOUTH, MAY 15.

INDEX-MAP.

MAY.

have but little difficulty in identifying most of the constellations and stars which were in more prominent positions in the corresponding diagram in April. These are Boötes, Coma Berenices, Virgo, Libra, and a few others. Looking due west, Cor Caroli is situated nearly half-way between the zenith and horizon, and easily distinguished by its being surrounded always by stars of comparatively small magnitude. The first star of importance from the zenith towards the south-west is Beta Boötis, between which and Arcturus are several bright objects; Arcturus itself being recognised by its ruddy appearance as well as by its superior lustre. Farther down in the same direction Spica can be seen. At this time a line drawn from the zenith through Arcturus passes through Spica. Near the south-west horizon, west of Spica, the stars in Corvus are still visible. Between Cor Caroli and the horizon all the principal stars in Leo can be identified, Regulus being very near to the horizon. This star is not included in the diagram, which contains only a few stars in the most easterly part of this constellation. The horizon from west to south contains parts of Sextans, Crater, Corvus, Hydra, Centaurus, and Lupus.

The complete list of constellations which adorn the southern half of the midnight sky of May is composed of Hercules, Corona Borealis, Boötes, Coma Berenices, Virgo, Libra, Serpens, Ophiuchus, Aquila, Sagitta, Delphinus, Equuleus, Vulpecula; and parts of Sextans, Crater, Corvus, Hydra, Centaurus, Lupus, Scorpio, Sagittarius, Capricornus, Aquarius, Cygnus, Lyra, Leo, and Canes Venatici.

We will now devote our attention to the sky north of the zenith, as illustrated by the view, looking north. The principal northern constellations visible at midnight on May 15th, are Ursa Major, Draco, Cygnus, Ursa Minor, Cepheus, and Cassiopeia, with portions of Leo, Cancer, Gemini, Auriga, Perseus, Andromeda, and Pegasus. First, let us, as usual, give a rapid glance on the meridian northwards, as far as the horizon. Passing over Kocab and Gamma Ursæ Minoris, the eye falls naturally upon Polaris. This standard star is still apparently in the same position as in preceding diagrams, though, to the astronomical observer, it has performed since January a part of a revolution in a small circle round the north pole, in a similar manner to the other stars. Polaris is really distant from the pole about one and a half degree; consequently its distance from the zenith or horizon differs about three degrees at its upper and lower transits over the meridian. For our purpose, however, Polaris is sufficiently near to the north celestial pole to serve as a zero-point which may be used as an easy reference for the identification of the other stars. The space between Polaris and the zenith, east and west of the meridian, is occupied entirely by Ursa Minor and Draco. The latter constellation completely separates the two Bears. In an old astronomical volume, printed at Venice in 1448, some very interesting figures of the constellations are inserted, including one in which the two Bears are completely enfolded in the embrace of Draco. Virgil, as rendered by Dryden, says :—

"Around our Poles the spiry Dragon glides,
And like a wandering stream the Bears divides."

Although several of the stars in the Lesser Bear are of small magnitude, its contour can be readily traced, Polaris being at the point of the tail, and Kocab and Gamma Ursæ

43

Minoris at the other extremity in the direction of the zenith, but slightly west of the meridian. These two stars are easily found, and by drawing a curved line from Kocab to Polaris the intermediate stars, though small, are distinctly visible. Directly north of the zenith, and a little west and east of the meridian, all the bright stars reaching in the west to the Great Bear, and in the east to the constellations Lyra and Cygnus, belong to Draco, the two brightest being Beta and Gamma, almost due east about twenty degrees. The principal portion of the sky west of Draco includes Ursa Major, below which are Leo Minor and Leo, Regulus being near the horizon north of west, but outside the limit of our diagram. The reader will perceive that all the seven principal stars in Ursa Major, Charles's Wain, are now in the northern half of the sky. He will also notice that the apparent position of this group of stars is very different to that in the first diagram of this series; for whereas the two pointers were in January approaching the upper meridian, they are now farthest from the zenith, and passing onward to the lower meridian below the pole. But their relative positions remain the same, and in whatever part of the sky they may be situated, Dubhe is always nearest to Polaris, and with Merak pointing out its position.

Below Ursa Major, in the north-west, is the bare constellation Lynx, and, near the horizon, Castor and Pollux may be seen on very fine nights. Between Polaris and the north horizon are Cameleopardalis and parts of Auriga and Perseus. The two bright stars, Capella and Beta Aurigæ, being circumpolar, may be observed near the horizon a little west of north.

The north-eastern part of the sky contains at midnight the constellations Cassiopeia, Cepheus, Andromeda, Cygnus, and Pegasus. Several bright stars in Cepheus may be noticed east of Polaris. Below Cepheus in the N.N.E. is Cassiopeia, with its well-known group called the Chair, and near the horizon numerous bright stars in Perseus and Andromeda can be distinguished on brilliant moonless nights. The principal objects in Pegasus can also be seen near the horizon in the E.N.E., while Alpha Cygni can be easily found in the same direction, about midway between the zenith and horizon. Vega, which has hitherto been a conspicuous star at midnight in this half of the sky, has passed over the boundary, and is now included in the diagram of the southern half.

JUNE.

IDNIGHT at midsummer, in the latitude of London, is so influenced by twilight, that many of the small stars, visible to the unassisted eye in the dark nights of winter, can only then be seen with telescopic aid, especially those north of the zenith. The sky near the north horizon is now more or less illuminated, while the general aspect of the heavens bears witness that there is no real night, but that there is constant day or twilight throughout the twenty-four hours. To those of our readers who are resident in the north of England, or in Scotland, the absence of complete darkness at midnight will be still more evident; but if we proceed to higher latitudes, or within the Arctic circle, we shall find that there will be no darkness at all, and that the phenomenon of the midnight Sun will at that hour be daily observed skirting the northern horizon. In London, however, there is always sufficient darkness on a midsummer midnight to observe stars down to the fifth magnitude with the naked eye, and consequently all contained in our diagrams.

Referring first to the diagram, looking south, or to the sky south of the zenith, it will be perceived that, although there is a general absence of very conspicuous constellations, yet several well-known stars are to be seen in different directions. Let us confine our attention at present to the sky east of the meridian, starting, as usual, from the zenith. The first star which naturally attracts our notice is Vega, about ten degrees south-east of that point. Very near Vega, in the same direction, are Beta and Gamma Lyræ, two stars of the third magnitude. Directly below these, and between Lyra and Aquila, are the small constellations Vulpecula, the Fox, and Sagitta, the Arrow. Aquila can be distinguished midway between the zenith and the horizon, by its group of three stars in the neck of the Eagle, the central and the largest being Alpha Aquilæ, or Altair. Between Aquila and the horizon, Capricornus is situated. A line drawn through the three chief stars in Aquila to the south-east horizon, passes nearly through Alpha and Beta Capricorni. The position of this sign of the zodiac is not otherwise well marked, owing to the paucity of large stars in that neighbourhood. North-east of Vega several bright stars in Cygnus are clearly visible, four of them being of the third magnitude. These are all generally known by a Greek letter, the star nearest to the zenith being Delta, the next

Gamma, then Epsilon, and the last Zeta Cygni. To the north of Gamma, Alpha Cygni, or Deneb, shines as a star of the first magnitude; but this object is included in the northern half of the sky, and consequently will be found in the diagram, looking north. Between Cygnus and the eastern horizon the space is occupied by the constellation Pegasus, one half of which at midnight is south, and the other half north of the imaginary line separating the two halves of the sky. Several bright stars in Pegasus can be seen near the horizon in the east. Three of these, together with the principal star in Andromeda, will form conspicuous objects in future diagrams, the combination being popularly known as the square of Pegasus. Between Aquila and Pegasus two small constellations, Equuleus and Delphinus, may be noticed, the latter more especially by a group of fourth and fifth magnitude stars. The horizon from due east to due west is occupied by several of the signs of the zodiac, the constellations, commencing from the east, being Aquarius, Capricornus, Sagittarius, Scorpio, Libra, and Virgo, the last-mentioned extending to a little north of west.

The principal stars on the meridian at this time are those in Ophiuchus, the chief object in which is Ras Alague, or Alpha Ophiuchi, about forty degrees from the zenith. Between Ophiuchus and the zenith the space is occupied solely by the constellation Hercules, which extends to a point very near the two bright stars in the zenith, Beta and Gamma in Draco. Ophiuchus spreads over a large portion of the sky on each side of the meridian, and reaches nearly to the south horizon. Excepting two or three stars near Ras Alague of the third magnitude, there is very little to attract the attention of observers in this constellation. West of the meridian, several well-known objects, the positions of which we have pointed out in the descriptions of the diagrams of preceding months, are still very conspicuous. First, near the horizon in the W.S.W., but out of the limits of our diagram, Spica, and other bright stars in Virgo, are on the point of setting. Arcturus, and a few other tolerably large objects in Boötes, are now a little south of west, about forty degrees from the horizon. They can be readily found by the ruddy appearance of Arcturus. Between Arcturus and the meridian, Alphecca and its companions, forming the Northern Crown, can be easily observed by the regularity and compactness of form of that small constellation. Directly south of Corona Borealis, and exactly midway between the zenith and horizon, Serpens, with a group of several bright objects, can be seen, the principal star being between the second and third magnitudes. This portion of the heavens, including Hercules, Serpens, and part of Ophiuchus, is peculiarly rich in stars of the second class. Near the S.S.W. horizon, the constellation Scorpio, with its bright star Antares, a few of the second and third magnitudes, and more than usual of the fourth and fifth, can now be easily recognised. Alpha and Beta Libræ are also visible, but they are within a short time of setting. Libra occupies the greater part of the south-western horizon.

The principal constellations in the south half of the midnight sky of London, in the middle of the month of June, may be briefly enumerated as follows:—Hercules, Lyra, Corona Borealis, Vulpecula, Sagitta, Delphinus, Equuleus, Aquila, Ophiuchus, Serpens,

THE MIDNIGHT SKY AT LONDON, LOOKING NORTH, JUNE 15.

INDEX-MAP.

THE MIDNIGHT SKY AT LONDON, LOOKING SOUTH, JUNE 15.

INDEX-MAP.

and Libra; and parts of Boötes, Cygnus, Pegasus, Coma Berenices, Virgo, Scorpio, Sagittarius, Capricornus, and Aquarius.

We will now transfer our attention briefly to the diagram, looking north, which is illustrative of the midsummer midnight sky of London north of the zenith. First, let us look directly overhead, where the two stars Beta and Gamma Draconis will be noticed as the brightest objects in their immediate neighbourhood, the more westerly star being Beta. Draco at this time occupies nearly the whole of the sky near the meridian between the zenith and Polaris, and all the stars as far as the Lesser Bear belong to that constellation. Commencing at Polaris, the form of Ursa Minor can be traced to the two bright stars, Kocab and Gamma Ursæ Minoris, which are a short distance west of the meridian in the direction of the zenith. The space between these stars and Charles's Wain, in Ursa Major, is also occupied by a part of Draco, which winds its way almost to the north side of Polaris. Confining our remarks at present to the stars west of the meridian, the principal constellations which fall under our view are, in addition to Draco and Ursa Minor, the whole of Ursa Major, Canes Venatici, Coma Berenices, Leo Minor, and Lynx, with portions of Boötes, Leo, and Auriga. The seven principal stars of Ursa Major now occupy the north-west sky, the pointers Dubhe and Merak being about midway between the zenith and horizon. Below Ursa Major, and near the horizon in the north-west, but out of the range of the diagram, portions of Leo and Leo Minor may be seen with some of the bright stars of Leo near the horizon. Looking due west from the two stars in Draco in the zenith, and passing down to the horizon, we traverse more or less through Hercules, Boötes, Canes Venatici, Coma Berenices, and Virgo, the last-mentioned being in the horizon. Below Ursa Major in the N.N.W. Lynx is situated, while in the north horizon the meridian divides the constellation Auriga, and its two bright stars, Capella and Beta Aurigæ, the former being slightly east, and the latter slightly west of due north.

The sky east of the meridian includes the whole of Cepheus, Cassiopeia, Andromeda, and Lacerta, and considerable portions of Perseus, Cygnus, Draco, Pegasus, Pisces, and Cameleopardalis. Let us now look due east from the two zenithal Draconian stars. At about one-third of the distance to the horizon, the eye will fall on Alpha Cygni, or Deneb, and near the horizon, on the bright stars of Pegasus. The stars of the latter constellation are now outside of the limit of this month's diagrams, but they will in future appear in the views of the south sky. The most attractive constellation in the north-east is Cassiopeia, which is about midway between the horizon and zenith, having Cepheus above, and Perseus and Andromeda below. The principal stars in Perseus, including Alpha Persei, and Algol, are visible near the N.N.E. horizon, east of the conspicuous stars Capella and Beta Aurigæ. Algol, although the second star in the constellation Perseus, is strictly in Medusa's head.

JULY.

———•———

THE principal constellations in the southern half of the midnight sky on July 15th, are Lyra, Aquila, Serpens, Ophiuchus, Capricornus, Aquarius, Pegasus, with considerable portions of Hercules, Cygnus, Corona Borealis, Pisces, and Cetus. Several first-class stars are now included in the lower diagram, such as Altair, Vega, Deneb, Ras Alague, and Markab. The immediate zenith is within the constellation Cygnus, whose chief star, Alpha Cygni, or Deneb, is now distant from that point by a few degrees only. Below Deneb are several bright stars of the third magnitude, most of which belong to Cygnus. The star nearest to Deneb, more southerly and slightly closer to the meridian, is Gamma Cygni, the next in order downwards, towards the south-east, is Epsilon Cygni, and the third, or lowest of the three, is Zeta Cygni. Pegasus is below Cygnus in the same direction, and lower still, reaching to the horizon, Aquarius is situated. If we draw an imaginary line from Deneb across Zeta Cygni, it will pass, at nearly the same distance, through Epsilon Pegasi, or Enif, a bright star between the second and third magnitudes. If this line be carried on a little farther, it points out Alpha Aquarii, or Sadalmelik, the principal star in the sign Aquarius. Between Pegasus and the meridian, the two small constellations, Equuleus and Delphinus, may be noticed, the latter being easily recognised by a group of moderately-sized stars. Looking due east, or rather south of east, at some distance below Deneb, or towards the left-hand of the upper part of the diagram, the eye will fall on several large stars in Pegasus, the nearest to the zenith being Beta Pegasi, or Scheat, and that nearest to the horizon Gamma Pegasi, or Algenib. Alpha Pegasi, or Markab, is the most southerly of these bright stars. These three objects, added to Alpha Andromedæ, or Alpherat, which, in this month, is included in the upper diagram, form the well-known group called the square of Pegasus. The actual horizon in the east is occupied by a part of Cetus. Capricornus is south-east of Aquila, approaching the meridian, but it contains no very prominent stars. On, or very near, the meridian, the chief stars worthy of notice are those composing the distinctive group in the breast or neck of the Eagle, Alpha, Beta, and Gamma Aquilæ. The central star of the three is Alpha, or Altair, the uppermost star is Gamma, and the lowest Beta, which is also the smallest. Other stars in Aquila can be noticed slightly west of the meridian. Beta Cygni, or Albiero, may be noticed between the Aquila group and the

52

THE MIDNIGHT SKY AT LONDON, LOOKING NORTH, JULY 15.

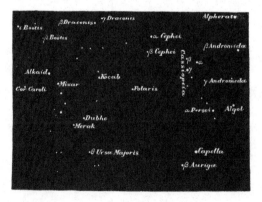

INDEX-MAP.

E

W

THE MIDNIGHT SKY AT LONDON, LOOKING SOUTH, JULY 15.

INDEX-MAP.

bright stars in Lyra. The intermediate space between Aquila and Beta Cygni is occupied by Vulpecula and Sagitta, two very small constellations. The sky near the south horizon contains portions of Piscis Australis and Sagittarius.

The western division of the south sky, a considerable portion of which will be below the horizon next month, contains the constellations Lyra, Ophiuchus, and Serpens, with parts of Hercules, Corona Borealis, Cygnus, Aquila, Sagittarius, Scorpio, and a few others. With the face directed to the south-west, the first object from the zenith is Vega, south of which are the two stars, Beta and Gamma Lyræ. About midway towards the horizon, we recognise Ras Alague, and the other principal stars in the constellation Ophiuchus, which extends nearly to the extreme limit of view. The horizon, south-west of Ophiuchus, contains several stars in Scorpio on the point of setting. Antares is invisible. Due west, below Vega, a considerable space is covered by Hercules; but a part of this constellation has passed into the northern half of the sky. Corona Borealis is also due west of the zenith: its principal star Alphecca, with three others of the semi-circular group, can be found on the right-hand side of the upper part of the diagram, on the point of passing into the northern half of the sky. The remaining members of the group, with several other stars in this constellation, have already passed into the northern half, and will consequently be found in the diagram, looking north. Below Hercules in the W.S.W., approaching the western horizon, all the stars in Serpens are still visible, also a few in Virgo, although they are within a half an hour of setting. On the whole, the south midnight sky in the middle of July is fairly attractive, excepting in the constellations south of the equator, where there is scarcely a star at this time above the third magnitude. This scarcity of brilliant objects in the lower part of the south sky is amply compensated by the brilliant appearance of the principal stars in Lyra, Cygnus, Aquila, and Pegasus, most of which are now in conspicuous positions.

In the middle of July at midnight, if the observer places himself so as to command a view of the heavens north of the zenith, he cannot fail to perceive the amount of change which has taken place in the positions of the principal stars, with respect to the meridian or horizon, since the date of the June diagrams. The zenith is now included within the constellation Cygnus; but no conspicuous star in the north sky is nearer than Beta and Gamma Draconis, which were the two zenithal stars in June. These are, however, still nearly overhead a short distance due west, Gamma being the nearer star to the zenith. By looking directly towards the Pole star, and therefore along the plane of the north meridian, we pass near the confines of Draco and Cepheus; but the stars in this direction are not attractive, or likely to receive attention, till we reach Polaris, which is as usual in its nearly constant position. Below Polaris as far as the north horizon, there is absolutely no star worthy of special notice, that part of the sky near the meridian being now occupied by Cameleopardalis and Lynx, both of which contain no star above the fourth magnitude, and very few above the fifth. But if, in July, the sky in the plane of the north meridian contains such a paucity of bright stars, the brilliant objects in other portions of the heavens, both east and west, will make up for the deficiency in that direction.

Taking the western side first we have in view nearly the whole of Draco, Boötes, Canes Venatici, Coma Berenices, Ursa Minor, and Ursa Major, and portions of Corona Borealis, Hercules, Leo Minor, and Lynx. Below Gamma and Beta Draconis, towards the W.N.W., a few of the chief stars in Hercules can be noticed, then those in Corona Borealis, beneath which is the greater part of Boötes, containing nearly all its principal stars, some of which, however, including Arcturus, are too near the western horizon to be included in the diagram. Nearly the whole of Ursa Minor is now west of Polaris. The space between the two Bears and almost to the seven chief stars of Ursa Major belongs to Draco. The position of the Ursa Major group is now on exactly the opposite side of the heavens to that exhibited in the January diagram, looking north. In that month Dubhe and Merak, the two leading stars in the waggon, are the nearest to the upper meridian, pointing in the direction of Polaris, with the shafts of the waggon directed to the north-east horizon. The heavens in July are viewed from the opposite part of the Earth's orbit to that in which the Earth was situated in January, and as we refer, by habit, our local time to the apparent motion of the Sun, the stars have all moved with respect to the Sun, in a direction from east to west, through a space equal to twelve hours of time. A brief explanation of the effect of the apparent motion of the Sun on the ecliptic from west to east was given in the description of the March diagrams. Dubhe and Merak, followed by the remaining stars, are now approaching the lower meridian. Their relative positions to Polaris always remain the same. The stars in the Ursa Major group have thus made a seasonal change of half a revolution round the Pole, independently of their diurnal rotation in the same direction. If we make a general comparison of the January and July diagrams of the north sky, this seasonal change in the positions of the stars can also be easily seen in all the other circumpolar constellations. This comparison of the views for successive months will familiarise the appearances of the different groups of stars, and enable the observer to become acquainted with the general form of the principal groups in each constellation. Ursa Major occupies about one-third of the north-west sky at midnight in July. Canes Venatici, with its star Cor Caroli, is visible to the left of Alkaid, the last star in the tail of the Great Bear. Coma Berenices is above the horizon below Cor Caroli, but its stars are too faint to be visible.

In the north-east sky, where all the stars are hourly rising towards the upper meridian, we find the constellations Cepheus, Cassiopeia, Perseus, Triangulum, and Cameleopardalis, and portions of Cygnus, Andromeda, Pisces, Aries, Taurus, Auriga, and Lynx. Cepheus is slightly east of the meridian above Cassiopeia's group, below which all the stars in Perseus are visible. In the N.N.E., near the horizon, Capella and Beta Aurigæ can always be seen in July at midnight. On looking due east, the eye falls on Alpha Andromedæ, or Alpherat, rather more than half-way towards the horizon, forming one of the "corner stones" of the square of Pegasus. In the E.N.E., Alpha and Beta Arietis, the two chief stars in Aries, are about twenty degrees above the horizon.

AUGUST.

A T midnight in the middle of August, the principal constellations on the meridian, south of the zenith, are Lacerta and Cygnus near that point, Pegasus, Aquarius, Capricornus, and Piscis Australis, the last-named being near the horizon. Their principal stars are, however, not at present on the meridian, excepting those in Aquarius. The most attractive objects in the southern half of the sky are the first-class stars, Vega, Alpha Cygni, Altair, and Alpha Ophiuchi, west of the meridian, and Fomalhaut, Markab, and Alpherat, east of the meridian.

We will commence our survey of the heavens for this month, by confining our attention at present to the sky south of the zenith. Let us then direct our attention as usual to the zenith, which is now occupied by the north-east corner of the constellation Cygnus. Lacerta and Cepheus join Cygnus near the zenith. Looking about due west, Alpha Cygni, or Deneb, strikes the eye about twelve degrees from that point, and farther on, in a due westerly direction, the brilliant Vega shines very conspicuously. South of Vega, but very near to it, Beta and Gamma Lyræ may be distinguished, and a little farther south, Albiero. In the diagrams for next month, most of these stars will have passed our imaginary line from due east to west, and will consequently appear in the view looking north. Below Vega, towards the west, there are still a few stars belonging to Hercules, including Rasalgeti, or Alpha Herculis, which, with Ras Alague, is visible about twenty degrees above the western horizon. Most of the preceding stars can be identified in the upper part of the right-hand side of the diagram, looking south. From the zenith to the western horizon we pass through three constellations only, Cygnus, Lyra, and Hercules, excepting a very small portion of Serpens in the horizon. In a south-westerly direction, below Cygnus, the stars in Aquila are the only objects of fair magnitude, although the constellations Vulpecula, Sagitta, Equuleus, Delphinus, and Sagittarius are passed over. Vulpecula and Sagitta are between Aquila and Cygnus, in the Milky Way. Delphinus and Equuleus are between Aquila and the meridian. The group of stars forming Delphinus is very plainly distinguished to the east of Altair. A line drawn from Vega, through the three chief stars in Aquila, and thence to the horizon, will pass over Alpha and Beta Capricorni. Capricornus extends from these stars as far as the meridian.

Now let us examine the division of the sky represented on the left-hand side of the

59

diagram, looking south. We see here many new stars and constellations which have scarcely come under our notice before. Those due east, or in the upper part of the left-hand side of the diagram, are Andromeda, Triangulum, Aries, and Cetus. South-east of the zenith, the four stars composing the square of Pegasus, Alpha, Beta, and Gamma Pegasi, and Alpha Andromedæ, can now be distinctly seen. Alpha Pegasi, or Markab, and Beta Pegasi, or Scheat, form the south-western and north-western corners of the square nearest to the meridian. Skirting the horizon from east to south several stars in Cetus and Piscis Australis are visible, including Menkar, or Alpha Ceti, in the east, Diphda, or Beta Ceti, in the south-east, and the bright star Fomalhaut a little east of south. The signs of the zodiac above the horizon at midnight in the middle of August are Sagittarius, Capricornus, Aquarius, Pisces, Aries, and parts of Taurus and Gemini, the last being near the horizon in the north-east.

That part of the sky immediately below Algenib, or Gamma Pegasi, is singularly devoid of large stars. It is principally occupied by the constellation Pisces, which, though of no great breadth, extends from the south near the meridian to almost due east. Pisces separates Pegasus and Andromeda from Cetus and Aries. There is no star below Gamma Pegasi, for some distance, greater than the fourth magnitude. Pisces can be identified in the diagram by several small stars east of Alpherat, or Alpha Andromedæ, and below Gamma Pegasi. The small asterism, Piscis Australis, is near the south horizon, below Aquarius and Capricornus. Its position is pointed out by its principal star Fomalhaut, which is a very conspicuous object in the south on clear autumn nights.

The Milky Way, which is generally brilliant at midnight in August, divides the heavens at this hour into two nearly equal portions. It can be easily traced from the zenith to the north-east and south-west horizons, passing through the constellations Cepheus, Cassiopœia, Perseus, Auriga, and Taurus, in the northern half of the sky, and through Cygnus, Vulpecula, Sagitta, and Aquila, in the southern half. When the Milky Way attains so elevated a position, the variable intensity of this luminous stratum is very clearly exhibited.

A first glance at the northern sky on August 15th at midnight, shows that the north-eastern portion is very barren, especially near the horizon. Below Cassiopeia and Perseus in the Milky Way, Capella and Beta Aurigæ are the only two stars likely to attract attention. Polaris is now the most easterly of all the stars in Ursa Minor, as Kocab and Gamma Ursæ Minoris are the most westerly. Looking along the plane of the meridian in the direction of Polaris, we can identify the constellations above and below the Pole. The space between the zenith and Polaris is very nearly occupied by Cepheus, whose two principal stars, Alpha and Beta Cephei, are west of, but not far distant from, the meridian. They can be easily distinguished from other stars near by their superior magnitude, Alpha being the nearest to the zenith. Gamma, the third star in Cepheus, is slightly east of the meridian, nearer Polaris. Passing over an unprolific part of Draco below Polaris, the sky as far as the horizon is occupied by a portion of Ursa Major, the pair of stars being Lambda and Mu Ursæ Majoris. The seven large stars in Charles's Wain are approaching the

THE MIDNIGHT SKY AT LONDON, LOOKING NORTH, AUGUST 15.

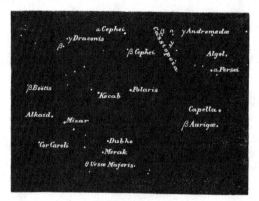

INDEX-MAP.

THE MIDNIGHT SKY AT LONDON, LOOKING SOUTH, AUGUST 15.

INDEX-MAP.

lower meridian, the two leading stars, Dubhe and Merak, being a little to the west of north. Below Eta Ursæ Majoris, at the end of the tail of the Great Bear, Cor Caroli can be perceived on very clear nights. Eta Ursæ Majoris is occasionally designated Alkaid, or Benetnasch, the Arabic denomination for the governor of the mourners, in allusion to the fanciful form of a bier and attendants, produced by the seven stars. All the stars in Ursa Minor are now west of the meridian, surrounded principally by the sinuous Dragon, whose chief stars can be recognised between Ursa Minor and Vega, the two bright stars Beta and Gamma Draconis being at the south-western extremity of Draco. About midway between Beta Draconis and Kocab, in Ursa Minor, is Zeta Draconis, of the third magnitude, the nearest large star to the Pole of the ecliptic. Low down in the western sky, most of the stars in Hercules can be noticed, and also the semicircular group of the Northern Crown. These are, however, outside the limits of the diagram. In the north-west, below Draco, a few stars in Boötes are still above the horizon, and also the greater part of Canes Venatici. The quarter of the heavens east of the meridian contains the whole of Cassiopeia, Perseus, Auriga, Cameleopardalis, and large portions of Cepheus, Ursa Major, Lynx, and Taurus. Cassiopeia is now very high in the heavens, approaching the meridian. Perseus is below Cassiopeia, and is easily distinguished by its principal stars Alpha Persei and Algol. Cameleopardalis is between Polaris and Auriga. Capella and Beta Aurigæ can be distinguished in the north-east. They are the most prominent objects in this portion of the north-eastern sky.

SEPTEMBER.

HE midnight sky of the middle of September is nearly of the same character as that which may be observed a month previously, excepting only that the stars have progressed about thirty degrees from east to west, and that a few of the stellar groups which give the winter sky so attractive an appearance are rising in the east. The chief constellations visible at this time in the southern half of the sky, from the zenith to the horizon, in the west, south, and east, are Aquila, Vulpecula, Delphinus, Equuleus, Capricornus, Aquarius, Pisces, Aries, Pegasus, Andromeda, Cetus, Triangulum, and portions of Cygnus, Lacerta, Taurus, Orion, Eridanus, Sculptor, and Piscis Australis. The chief stars are Altair, Aldebaran, Fomalhaut, and the four stars composing the square of Pegasus.

Cassiopeia occupies the zenith, but south of that point there are no large stars for several degrees. Looking due south, we shall find that the stars in the square of Pegasus are now at their highest point, Beta Pegasi and Markab being slightly west, and Alpherat and Algenib east of the meridian. West of the zenith, the stars in Lacerta, Cygnus, Vulpecula, and Aquila, can be noticed, the three chief stars in Aquila being in the W.S.W., and on the right-hand side of the diagram. In a south-westerly direction, beginning at the zenith, we pass over a part of Andromeda, Pegasus, Aquarius, and Capricornus, and very near the horizon in the S.S.W., the position of Piscis Australis, indicated by its principal star, Fomalhaut, can be found. The sky on this side of the meridian is not very brilliant at this time, although there are a large number of stars of the third and fourth magnitudes. If we now consider the eastern sky, which will soon be so rich in conspicuous objects, we shall find that the major part of Andromeda is east of the meridian, including all its principal stars. Alpha Andromedæ, or Alpherat, being the north-eastern star of the square of Pegasus, can readily be distinguished, Beta is a little to the north-east, and Gamma a short distance farther in the same direction. By prolonging a line drawn from Markab to Alpherat, it will pass through Beta and Gamma, and farther on it will meet Alpha Persei. If these three stars be joined on to the square of Pegasus it will be perceived that their combined form bears some resemblance to Charles's Wain in Ursa Major. Gamma Pegasi, or Algenib, is the south-eastern star of the square. South-east of Andromeda, and about half-way between the zenith and horizon, Aries, the first of the zodiacal signs, can be

W   E

THE MIDNIGHT SKY AT LONDON, LOOKING NORTH, SEPTEMBER 15.

INDEX-MAP.

THE MIDNIGHT SKY AT LONDON, LOOKING SOUTH, SEPTEMBER 15.

INDEX-MAP.

identified by a group of tolerably bright stars, of which Alpha and Beta, near together, are the principal. Beta is the nearer of the two to the meridian. Pisces cannot boast of any star greater than the third and a half magnitude; but it extends, nevertheless, over a considerable portion of the heavens, from near Beta Andromedæ to below Markab, the south-westerly star in the square of Pegasus. In the diagram the general form of the constellation can be seen by a kind of curve of small stars in the direction just indicated. Cetus also spreads over a large part of the south-eastern sky, and its position is pointed out by several stars of the second and third magnitudes. Menkar, or Alpha Ceti, is in the E.S.E., at some distance below Aries, while Beta Ceti is in the S.S.E. Sculptor is below Cetus near the meridian, and Eridanus occupies the south-east horizon. East of Aries and Cetus, Taurus, the second zodiacal sign, can be recognised by its bright star Aldebaran, between which and the zenith, the well-known group of the Pleiades can be seen. Aldebaran is out of the range of the diagram, but the Pleiades are inserted almost due east of the zenith at the left hand. The eastern horizon is occupied by Orion, but the principal stars by which this constellation is generally distinguished are still below the horizon on September 15th. The small constellation Triangulum is situated between Andromeda and Aries, its principal star being nearly midway between Alpha Arietis and Beta Andromedæ. Near Triangulum, a small portion of Perseus is now included in the south map, but all its chief stars are still contained in the diagram, looking north. The Milky Way now passes through the zenith, from almost due east to west.

The sky north of the zenith now contains the two bright stars Vega and Capella on opposite sides of the meridian. Vega can be recognised on the left side of the diagram, looking north, and Capella in a corresponding position on the right side. Cassiopeia is approaching the upper meridian near the zenith, followed by Perseus, whose principal stars are still strictly in the northern half of the sky. Auriga is in the E.N.E. under Perseus, its position being indicated by Capella and Beta Aurigæ. In the same direction, Castor and Pollux are shining near the horizon. Castor just appears in the diagram, but Pollux is not yet within its limits. Between Capella and Polaris nearly the whole space is occupied by Cameleopardalis. This part of the heavens is devoid of large stars. Turning our attention to the constellations west of the meridian, bright objects down to the third magnitude are profusely scattered about in all directions, most of them belonging to important asterisms. Let us commence our survey with Cygnus, most of whose stars have passed since last month from the southern to the northern map. Deneb, or Alpha Cygni, is near the upper part of the map, or almost due west in the heavens. In the W.N.W., Vega points out the position of Lyra, below which, towards the horizon, the stars generally belong to Hercules. Again, starting from the zenith, but this time towards the north-west, the constellation Cepheus fills up the space for more than thirty degrees. All its principal stars, and even its general shape, can be recognised with very little trouble. The reader is referred to the description of this constellation for the relative positions of the different members. Cepheus separates Cassiopeia from Ursa Minor and Draco. Below Cepheus, all the stars in the central north-west sky are constituents of

Draco, which extends from Lyra to the eastern side of Ursa Minor. Kocab and Gamma Ursæ Minoris, in the breast of the Lesser Bear, are N.N.W. of Polaris. These stars are sometimes termed the guards or wardens of the pole. A few stars in Boötes may be seen very near the N.N.W. horizon under Draco.

Ursa Major is in a very attractive position, although its altitude is low. It is now exhibited in the natural form of a waggon or plough, and all its stars can be viewed in great perfection when that part of the heavens is free from haze. Our imaginative conceptions of the form and arrangement of this celebrated collection of stars have been gathered chiefly when Ursa Major is at a comparatively low altitude in the north-west, north, or north-east. When this group of stars is on the upper meridian, it passes so nearly overhead that, in ordinary cases, it is not included in the field of vision without looking upwards to the zenith, an act which is seldom performed without a special effort. Dubhe and Merak, the Pointers, have passed the lower meridian from west to east at midnight on September 15th, Gamma Ursæ Majoris is on the meridian, and Alkaid, or Benetnasch, which marks the end of the Bear's tail, is still considerably west of that point.

OCTOBER.

A T midnight in October, the eastern and south-eastern sky is becoming enriched by some of the most conspicuous constellations, including Orion, Taurus, and Gemini, but, excepting Taurus, they are outside the limit of the diagrams. Andromeda, Perseus, and Cassiopeia now meet in the zenith, near which several important stars are visible. Let our attention be first directed to the diagram, looking south. The stars situated a few degrees east of the zenith belong to Perseus; Algol, or Beta Persei, in Medusa's head, being near the south edge of the Milky Way. Proceeding downwards towards the eastern horizon, or the upper part of the left-hand side of the diagram, we pass over the eastern corner of Taurus, and the second star in that constellation, Nath, or Beta Tauri. East of this star, but too near the horizon to be included in the diagram, several stars in Gemini and Canis Minor are visible, Procyon having just risen due east. Confining ourselves to this portion of the heavens for the present, or rather in an E.S.E. direction, the splendid group of first and second magnitude stars in Orion, which add so much lustre to the winter sky of the northern hemisphere, can be seen without any special instruction. Rigel and Bellatrix, the two westerly stars of the Quadrilateral, are inserted in the diagram. Between Algol and Orion, but nearer the latter, Aldebaran and its companion stars, the Hyades, are easily recognised. A line drawn from the zenith through Algol to the three stars in the belt of Orion, passes through the Hyades. The universally-known Pleiades group can be seen to the right of Aldebaran. In the south-east, most of the sky is occupied by Eridanus, an extensive constellation; but as it contains no star greater than the third magnitude visible at London, this portion of the heavens appears bare in comparison with that to which we have just drawn the reader's attention. Eridanus extends from near Rigel, the south-west star in the quadrilateral of Orion, to within a few degrees of the south meridian near the horizon.

The following constellations are now on or near the south meridian, beginning at the zenith: Andromeda, Triangulum, Pisces, Aries, Cetus, and Sculptor, the last mentioned being in the horizon. Although there are but few large stars now on the meridian, those belonging to the separate constellations can be pointed out easily in the diagram, if we recollect that a line drawn from the zenith through the exact centre as far as the south horizon will represent the corresponding position of the celestial meridian, looking south.

73

Not very far from the zenith, a star rather brighter than those near, and slightly left of the meridian line, is Gamma Andromedæ, a beautiful triple star. One to the right of the meridian is Mirach, or Beta Andromedæ. A little lower, above two conspicuous stars, Alpha Trianguli can be noticed, the two large stars being respectively Alpha and Beta Arietis. To the right of these a line of small stars in Pisces may be seen, marking the position of the two fishes and their connecting ribbon. Below Pisces a few stars of the third and fourth magnitudes belong to Cetus.

West of the meridian, the principal constellations above the horizon at midnight are Pegasus, Equuleus, Pisces, Aquarius, and portions of Andromeda, Cetus, and Sculptor. The chief stars on this side of the meridian are those in Andromeda and Pegasus. That near the zenith, a little to the right of the meridian, we have already pointed out as Beta Andromedæ. Between it and Alpherat is Delta Andromedæ, of the third magnitude. The stars composing the celebrated square of Pegasus have been explained in preceding months; it is enough, therefore, to say at present that Alpherat, or Alpha Andromedæ, is the nearest of the four to the zenith, Beta Pegasi is in the north-western corner of the square, Alpha Pegasi, or Markab, in the south-western, and Gamma Pegasi in the south-eastern. A tolerably bright star to the right of the square, and nearly at the limit of the diagram, is Epsilon Pegasi. That south-west of Markab is Zeta Pegasi, and in the same direction two others, near the right-hand limit of the diagram, are Gamma and Alpha Aquarii. The only object of large magnitude south of these stars is Beta Ceti, which may be noticed towards the lower part of the diagram on the right-hand side of the meridian.

Cassiopeia is now situated near the zenith, its principal stars having just passed the meridian. In the diagram of the north sky they are near the upper boundary line, forming some resemblance to an antique chair, of which the lower portion is composed of Alpha, Beta, Gamma, and a smaller star, Kappa. Beta is the most westerly of all, and Alpha the most southerly. To see Cassiopeia in the heavens at midnight in October, the observer's face being directed towards the north, it is best to find first the position of the polar star, and then gradually to look upwards, almost to a point overhead. Between Polaris and Cassiopeia, there are no objects of sufficient magnitude to attract attention, excepting perhaps Gamma Cephei, which would be almost touched by a line drawn from Beta Cassiopeiæ to Polaris. In the north-western quadrant of the sky, several stars of large magnitude are visible in different directions, including Vega, Kocab, and Deneb. From the zenith to the north-western horizon, we pass through Cassiopeia, Cepheus, Cygnus, Draco, Lyra, and Hercules. Looking due west, or along the upper boundary line of the diagram, there is no star sufficiently bright to notice especially, but towards the W.N.W., following the course of the Milky Way, several stars in Cygnus may be pointed out. The first is Deneb, or Alpha Cygni. That west, or to the left of Deneb, is Gamma Cygni, in the Milky Way. In the other arm of the Milky Way Epsilon Cygni may be seen. Zeta and Delta are apparently above and below the Milky Way respectively. North of Cygnus, the position of Lyra can be recognised by Vega, and the two neighbouring

74

THE MIDNIGHT SKY AT LONDON, LOOKING NORTH, OCTOBER 15.

INDEX-MAP.

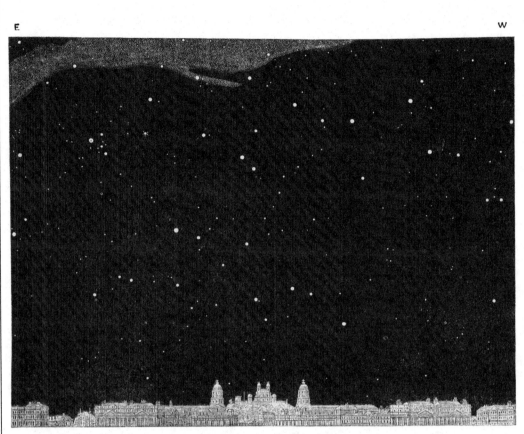

THE MIDNIGHT SKY AT LONDON, LOOKING SOUTH, OCTOBER 15.

INDEX-MAP.

stars, Beta and Gamma Lyræ. To the right of Vega, and near the north meridian, Gamma and Beta Draconis are conspicuous, and below these several stars in Hercules are near the horizon. Below Cassiopeia, and between Cygnus and the meridian, Cepheus is situated. The two principal stars in Cepheus can be found by drawing a line from Alpha Cygni to Polaris, Alpha Cephei being the nearer one to Cygnus. All the chief stars in Ursa Minor are now to be found between Polaris and the north horizon. Kocab and Gamma Ursa Minoris at one end, and Polaris at the other, define the extent of this small but important constellation. Between Kocab and the horizon, the sky is occupied by portions of Draco and Boötes. Alkaid, the last star in Ursa Major, is exactly on the meridian.

East of the meridian the principal stars above the horizon are Capella, Castor, Pollux, and those in Ursa Major. With the exception of Ursa Major, all the chief stars in this quarter of the sky are in the E.N.E., and mostly in Perseus, Auriga, and Gemini. These three constellations are now partly in the northern, and partly in the southern half of the sky. Due east from the zenith, the nearest large star is Alpha Persei. The next conspicuous object is Capella, followed by Beta Aurigæ, and below these Castor and Pollux may be easily recognised in the heavens, Castor being the upper star of the two. Castor is inserted at the right hand limit of the diagram. The extensive district between Castor and Polaris, from east to west, and from Capella to Charles's Wain, in Ursa Major, from south to north, is occupied principally by the constellations Lynx and Cameleopardalis. In the north-east, below Lynx, Leo Minor and the northern portion of Leo have just risen. The chief part of Cancer is also above the horizon, between Leo and Gemini.

NOVEMBER.

HE midnight sky in the middle of November contains a large instalment of bright winter stars, with which we are so familiar in the evenings of the first months of each year. In the diagram looking south, many of these can probably be identified by the reader at a glance, or by a reference to the corresponding index-map. The zenith is now occupied by the constellation Perseus, the chief star of which, Alpha Persei, is very near that point slightly west of the meridian. Algol is a little lower towards the S.S.W. Looking due west, below Perseus, the eye passes over Andromeda and Pegasus, in which the four stars in the square of Pegasus may be recognised. Of these, Alpherat, or Alpha Andromedæ, and Gamma Pegasi, may be noticed in the right-hand upper corner of the diagram. If we look in a south-westerly direction, below the two bright stars in Perseus, the next conspicuous objects are Alpha and Beta Arietis; and lower down, nearly to the horizon, most of the visible stars are included in the constellation Cetus. Nearly all these are inserted in the diagram, reaching from a short distance west of the meridian to the south-west limit of the map. The most easterly star in Cetus is Menkar. Proceeding onwards towards the south-west, the next star is Gamma Ceti, preceded at some distance by Mira Ceti, a most curious variable star. Beta Ceti is near the limit of the diagram, at the right-hand side. It will not be difficult to discover most of the chief stars in Cetus, as, although it is a large constellation containing numerous small stars, yet the principal objects are well separated from each other.

The Pleiades are now on the meridian, about twenty-eight degrees from the zenith; these popular stars can always be distinguished without difficulty, by their nebulous or cloudy appearance to the naked eye. The central and principal star of the group, Alcyone, is about the third magnitude. At this moment only three constellations are on the meridian, Perseus, Taurus, and Eridanus. Alpha Persei has just passed the meridian near the zenith. Perseus occupies most of the sky between the zenith and the Pleiades. The star of the third magnitude near Alpha Persei is Delta, the two between it and the Pleiades are Epsilon and Gamma Persei. Algol, in a narrow branch of the Milky Way, is due west of Epsilon Persei. Eridanus, joining Taurus on the meridian, is a very extensive asterism, and many modern unsuccessful attempts have been made to reduce it by the formation of several small constellations out of some of its outlying portions. The stars

W

E

THE MIDNIGHT SKY AT LONDON, LOOKING NORTH, NOVEMBER 15.

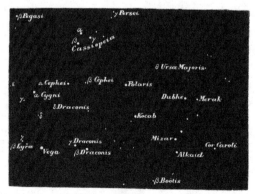

INDEX-MAP.

THE MIDNIGHT SKY AT LONDON, LOOKING SOUTH, NOVEMBER 15.

INDEX-MAP.

near the meridian, in the lower portion of the diagram, all belong to Eridanus; none of them are, however, greater than the third magnitude. Achernar, of the first magnitude, is not only the principal star in Eridanus, but also one of the brightest in the southern hemisphere. It never rises, however, above the horizon of London.

Let us now direct our attention to the sky east and south-east of the meridian. Here we have at one view a perfect galaxy of stars, including those in Taurus, Auriga, and, lower down half-way between the zenith and the eastern horizon, Castor and Pollux. These stars can be recognised near the upper boundary line on the left-hand side of the lower diagram. Near the horizon, but outside our limit, Regulus and the stars in Leo add to the beauty of the eastern midnight sky of this month. From the meridian to the E.S.E. horizon, we pass above Orion and through Taurus, Gemini, Canis Minor, and Hydra, and near the first-class stars Aldebaran and its companions the Hyades, Beta Tauri, and Procyon. The position of Orion in the heavens requires no explanation, as that is always pointed out by the well-known three stars in the Warrior's belt. The two upper stars are Betelgeuse in the north-east and Bellatrix in the north-west corner of the quadrilateral, Rigel being in the south-west corner. Sirius, the brightest of the fixed stars, is near the horizon in the south-east. A tolerably bright star to the right of Sirius is Beta Canis Majoris. A few objects below the quadrilateral of Orion belong to Lepus. Between Sirius and Procyon the sky is occupied by Monoceros, a district void of stars above the fifth magnitude. Near the horizon in the S.S.E., one or two objects in Canis Major below Sirius, and in Columba, can be seen at midnight.

In addition to the usual circumpolar constellations, the northern sky contains, at midnight, portions of Pegasus, Cygnus, Andromeda, Vulpecula, Lyra, Hercules, Boötes, Canes Venatici, Coma Berenices, Lynx, Leo Minor, and Leo. The diagram, looking north, contains several stars in some of these constellations. If we examine the north-western quadrant first, the bright star-group of Cassiopeia will at once attract our notice in the W.N.W., at no great distance from the zenith. Below Cassiopeia, in a north-westerly direction, the three principal stars in Cepheus may be recognised as forming an arc of a circle, of which Beta, the lowest of the stars in Cassiopeia, is nearly the centre. Between Cepheus and the horizon, the sky is occupied by Cygnus, whose chief star, Deneb, or Alpha Cygni, is a prominent object twenty-five degrees high. Vega is near the horizon in the N.N.W. The two bright stars between Vega and the meridian, are Gamma and Beta Draconis, Gamma being that nearest to Vega. Draco now incloses Ursa Minor on the west, north, and east. Kocab, Gamma Ursæ Minoris, and the small stars leading to Polaris, are all to the north of the pole. A considerable portion of the north-east sky is occupied by Ursa Major, which extends some distance south and east of Charles's Wain. The pointers, Dubhe and Merak, are the uppermost stars of the group, the third is Gamma Ursæ Majoris, the fourth Delta, the fifth Epsilon, or Alioth, the sixth Zeta, or Mizar, which is double, and the seventh Eta, or, as it is sometimes called, Alkaid, or Benetnasch. This last star is fourteen degrees above the horizon. The companion star of Mizar is named Alcor, and can be perceived occasionally by the naked eye, and very clearly with an opera glass, or small hand-telescope. South of Ursa Major and Draco, the large space reaching to the zenith is chiefly occupied by Lynx and Cameleopardalis, two constellations without a single large star. In the eastern horizon, but outside the limit of the diagram, several bright stars in Leo have just risen. The actual horizon from west to north and east is occupied by Pegasus, Vulpecula, Cygnus, Lyra, and Hercules, west of the meridian, and Boötes, Canes Venatici, Coma Berenices, and Leo, east of the meridian. Cor Caroli, the principal star in Canes Venatici, is visible east of Eta Ursæ Majoris. Boötes and Hercules are in the north horizon, the former being slightly east of the meridian, and the latter west of the meridian; the two constellations are separated by Corona Borealis. The position of these three asterisms is, however, much too low, at this time, for any of their stars to be seen with advantage.

DECEMBER.

THE midnight sky is perhaps more brilliant in December than in any other month of the year at the same hour. Some of the finest constellations which adorn the heavens south of the zenith are now in conspicuous positions. Among these, Aries and Taurus, west of the meridian, Auriga, Orion, and Canis Major, almost due south, or on the meridian, and Canis Minor, Gemini, and Leo, in the east, may be specially mentioned. Adding to the above the circumpolar constellations north of the zenith, Ursa Major, Ursa Minor, Draco, Cassiopeia, and others, we have at one view the majority of the principal stars visible in the northern hemisphere. Any one stationed so as to command the whole of the sky above the horizon, can at this time perceive the following first-class stars :—Aldebaran, Betelgeuse, Rigel, Delta, Epsilon, and Zeta Orionis (the three stars in the belt of Orion), Capella, Sirius, Procyon, Castor, Pollux, Regulus, Denebola, Alpherat, Vega, Deneb, and the stars composing the well-known groups of Ursa Major, Perseus, and Cassiopeia.

The magnificent appearance of the south meridian midnight sky of December is, however, principally due to the prolific region occupied by the constellations Taurus and Orion. Aldebaran, the chief star in Taurus, is exactly midway between Bellatrix and the Pleiades. The vicinity of Aldebaran is unusually rich in small objects, and includes the group of the Hyades. This brilliant assemblage of stars has been thus eulogised by the astronomical rhymester :—

"In lustrous dignity aloft, see Alpha Tauri shine,
The splendid zone he decorates, attests the Power divine :
For mark around what glitt'ring orbs attract the wandering eye,
You'll soon confess no other star has such attendants nigh."

Beginning with the south-western quarter of the sky, or the right-hand side of the diagram, looking south, the observer is requested to look over-head, when he will notice the very bright star, Capella, about eight degrees from the zenith. Due west of Capella, the Perseus group can be distinguished in the Milky Way; its principal star, Alpha Persei, will be found inserted in the diagram, looking north, while Algol is in that representing the sky south of the zenith; the imaginary line from due east to due west consequently passes between these stars, thus dividing the constellation Perseus. Alpha and Beta

87

Arietis, the chief stars in Aries, are below Perseus towards the west; they can be identified in the diagram near the right-hand upper corner. Between Aries and the western horizon, the space is wholly occupied by Pisces. Taking Capella as a zero-point, and looking towards the south-west, we pass over the thickly-studded constellation Taurus, the position of which is easily recognised by the Pleiades group, and by Aldebaran, with its companion stars the Hyades. Below Taurus and Aries, in this direction, Eridanus and Cetus extend to the horizon. The stars immediately below Aries belong to Cetus, and those near the south-west horizon to Eridanus.

From the zenith to the horizon, along the plane of the meridian, we pass over Auriga, which now occupies the sky directly overhead and several degrees south. Its principal stars, Capella and Beta Aurigæ, are near the zenith, the latter being on the meridian. Below Auriga, the two signs of the zodiac Taurus and Gemini join each other, and beneath these the brilliant assemblage of stars composing the Orion group is at its greatest elevation. Directly below Orion, the small constellation Lepus can be identified by some moderately bright stars; and south of Lepus, very near to the horizon, a few stars in Columba can be seen on clear nights when the south horizon is free from haze. The bright star about half-way between the zenith and the upper stars in the quadrilateral of Orion is Beta Tauri, or Nath, the second star in Taurus. In the notes on the constellations, we have given a separate diagram of Orion on a larger scale than that adopted in the sky views, and have also inserted the names of all the principal stars in that favourite constellation by their usual Arabic or Greek designation. For this reason it will be sufficient to state here that the north-western star of the quadrilateral is Bellatrix, or Gamma Orionis; that in the north-east is Betelgeuse, or Alpha Orionis; that in the south-west corner is Rigel, or Beta Orionis; and that in the south-east is Kappa Orionis. The most westerly of the three stars in the belt is Mintaka, or Delta Orionis; the central one is Alnilam, or Epsilon Orionis; and the most easterly star is Zeta Orionis, or Alnitak.

Some very prominent stars are contained in the south-eastern quarter of the sky, but still a considerable portion of this division of the heavens is comparatively bare, especially east of Orion and Canis Major. Sirius, the most conspicuous of all the fixed stars, and several other objects in Canis Major, are now visible in the S.S.E.; some of them are, however, near the horizon. At some distance east of Betelgeuse, after passing across the Milky Way, Procyon can be detected as much from its intrinsic lustre, as by its forming, with Betelgeuse and Sirius, the most splendid stellar equilateral triangle in the heavens. North of Procyon, the twin stars Castor and Pollux can be recognised at a glance. The tolerably bright objects between Pollux and Orion all belong to Gemini. Looking due east all the stars in Leo are distinctly visible; many of them are, however, very near the eastern horizon. In the left-hand upper corner of the diagram, looking south, Regulus, Gamma Leonis, and the remaining stars in the Sickle are inserted, but the other stars in Leo are outside the limit of the diagram. Although the absence of large stars in the south-east makes that part of the heavens look, by mere contrast, almost bare, yet one scarcely notices the defect on the clearest winter's night, on account of the unusual brilliancy of the meridian

THE MIDNIGHT SKY AT LONDON, LOOKING NORTH, DECEMBER 15.

γ Andromedæ • α Persei

β Andromedæ

• Alpherat

α. γ.
β. Cassiopeia

⊖ Ursæ Majoris.

Dubhe. • Merak

• Polaris

β Cephei.

Kocab.

• Mizar

α Cephei.

Alkaid •

• α Cygni

γ Draconis • β Draconis

INDEX-MAP.

THE MIDNIGHT SKY AT LONDON, LOOKING SOUTH, DECEMBER 15.

INDEX-MAP.

sky. Excepting Alpha Hydræ, or Alphard, there is scarcely an object between Canis Minor and the horizon above the fourth magnitude. Alpha Hydræ can be detected in this thinly-studded region of the heavens by this total absence of any star of equal magnitude. The constellations Monoceros, between Orion and Canis Minor, Sextans, below Leo, and Cancer, between Gemini and Leo, contain no star of sufficient prominence for a special indication of its position.

North of the zenith, the midnight sky contains portions of several well-known constellations, which at the same hour in other seasons of the year are included in the south sky. Many of them are, however, near the horizon. Beginning at the west, and passing round the horizon from west to north and east, some of the principal stars in Pegasus, Cygnus, and Lyra, can be seen very low down on the western side of the meridian, and those in Hercules, Boötes, Coma Berenices, and Virgo, on the eastern side. The horizon due west and east is occupied by Pisces and Virgo respectively. Nearly the whole of Andromeda has passed from the diagram, looking south, to that looking north, since last month. Polaris is very easily identified, as no other important star is situated between the zenith and the pole. This district of the heavens is entirely occupied by Cameleopardalis. The meridian between Ursa Minor and the north horizon passes through the widest portion of Draco, whose two brightest stars, Beta and Gamma, are at about twelve degrees altitude, Beta having just passed the lower meridian, to which Gamma is approaching. These two stars indicate the position of the head of the Dragon.

In the north-western sky, the chief stars in Cassiopeia, Cepheus, and Andromeda, can all be readily traced. Cassiopeia is very favourably situated in the north-west, midway between the zenith and horizon. The apparently lowest object of this group is Beta Cassiopeiæ, the leading star of the constellation. In the W.N.W., or in the upper part of the diagram on the left hand, the stars all belong to Perseus or Andromeda. It has been already mentioned that one portion of Perseus is contained in the northern half of the sky, and the other in the southern half. Below Perseus, towards the west, the first tolerably-bright star is Gamma Andromedæ, the next Mirach, or Beta Andromedæ, and that near the limit of the diagram Alpherat. All these stars can be recognised in the heavens, in the order we have given, by looking a little south of west. Cepheus is north of Cassiopeia, and north-west of Polaris, and in the same direction, near the horizon, Deneb, or Alpha Cygni, is the brightest star now visible in this quarter of the sky. Vega is above the horizon, but only by two-thirds of a degree. The haze is, however, always dense enough to obscure it. This star has been observed occasionally in former years, when in this position, through the Greenwich meridional instruments; but since the growth of London in the north-eastern suburbs, the increased impurity of the atmosphere has prevented any observations being made.

Ursa Major is the principal constellation in the north-eastern quarter of the sky. Its position can be perceived by its seven chief stars. It extends to within twenty degrees of the zenith. Dubhe and Merak, the pointers, are now the uppermost of the stars in Charles's Wain. This group is approaching that part of the sky in which they were situated in our first diagram in January. By an inspection of the twelve diagrams of the north sky, the effect of the seasonal changes, independently of the daily variations owing to the diurnal rotation of the earth, can be easily followed by noticing the monthly positions of this well-known group. Every day in the year these stars, in common with all the others, have apparently revolved in a circle, of which the celestial pole is the centre. They have also, as we explained in March, revolved around the pole once during the year, as a consequence of the revolution of our globe in its orbit around the Sun in the same time. They have, in fact, apparently revolved 366 times in 365 days. All the circumpolar constellations would serve as examples of the seasonal variations of the positions of the stars, for any given hour, as well as Ursa Major, but the well-defined form of this group makes it more easily recognised than any other. South of the principal group in Ursa Major, or near the upper part of the right-hand side of the diagram, a close pair of stars, Iota and Kappa Ursæ Majoris, situated in the Great Bear's right fore-foot, may be noticed, and a similar pair a little lower, but more to the right, are Lambda and Mu Ursæ Majoris, in the Bear's right hind-foot. If we except Ursa Major, there are not many bright stars in December in the north-eastern midnight sky. Kocab and Gamma Ursæ Minoris, to the left of Charles's Wain, Cor Caroli to the right, and a few stars in Draco, below Ursa Minor, are the principal objects after the seven in Ursa Major.

THE

MIDNIGHT SKY

OF THE

SOUTHERN HEMISPHERE.

" Go, take the wings of morn,
    And fly beyond the utmost sea ;
Thou shalt not find thyself forlorn,
    Thy God is still with thee :
And where his Spirit bids thee dwell,
There, and there only, thou art well.

" Soon the wide world, between
    Our feet, conglobes its solid mass ;
Soon lands and waters intervene,
    Which I must never pass :
Soon day and night with thee are changed,
Seasons reversed, and clime estranged.

" When tropic gloom returns,
    Mark what new stars their vigils keep.
How glares the *Wolf*, the *Phœnix* burns ;
    And on a stormless deep,
The *Ship* of heaven, the patriarch's *Dove*,
The emblem of redeeming love. *

" While these enchant thine eye,
    Oh think how often we have walked—
Gazed on the glories of *our* sky ;
    Of higher glories talked,
Till our hearts caught a kindling ray,
And burned within us by the way."

JAMES MONTGOMERY.

* The Southern Cross.

FEBRUARY.

THE celebrated traveller Alexander Von Humboldt, has remarked that "one experiences an indescribable sensation when, as he approaches the equator, and especially in passing from one hemisphere to the other, he sees the stars with which he has been familiar from infancy gradually approach the horizon and finally disappear. Nothing impresses more vividly on the mind of the traveller the vast distance to which he has been removed from his native country than the sight of a new firmament. The grouping of the larger stars, the scattered nebulæ rivalling in lustre the Milky Way, and spaces remarkable for their extreme darkness, give the southern heavens a peculiar aspect. The sight even strikes the imagination of those who, although ignorant of astronomy, find pleasure in contemplating the celestial vault, as one admires a fine landscape or a majestic site. Without being a botanist, the traveller knows the torrid zone by the mere sight of its vegetation; and without the possession of astronomical knowledge, perceives that he is not in Europe, when he sees rising in the horizon the great constellation of the Ship, or the phosphorescent clouds of Magellan. In the equinoctial regions, the earth, the sky, and all their garniture, assume an exotic character." Experiences similar to those of the great philosophical traveller are doubtless felt by most intelligent voyagers, who for the first time find themselves in view of the gorgeous grouping of some of the principal southern constellations. A writer in the *Leisure Hour*, after alluding to the changes in one's habits and feelings produced by the great difference of temperature between the banks of the Avon and the tropics, says: "The Moon, too, only appeared brighter, fairer, and better defined through a more transparent atmosphere. But the stars were different. Walking the deck on fine nights, the heavens seemed palpably changed; and the thought of being far, far from home, was impressed upon the mind with a power never known before. Stars, which had been watched in the northern sky with interest and delight in the days of childhood and youth, drooped towards the horizon, and were at length looked for in vain. Others customarily seen towards the south were high overhead, while strangers appeared in the direction we were pursuing, ascending higher and higher till there was almost a new heaven aloft, without any intimation that the old earth had passed away."

The reputation which the sky of the southern hemisphere has obtained for its superior brilliancy over that of the northern hemisphere, depends principally upon three or four rich constellations which are situated at no great distance from the most splendid portion of the Milky Way. In other directions, especially within twenty degrees around the south pole, the heavens are particularly devoid of large stars. Taken as a whole, the circumpolar stars visible at London, and which are consequently invisible, for the most part, in our southern colonies, can vie in magnitude and interesting grouping with the larger portion of the southern heavens which never rise above the horizon of London. But the more than usual brilliancy of the Via Lactea in the neighbourhood of the Centaur and the Southern Cross, coupled with the profuse collection of large stars in these two constellations and the Ship, makes this portion of the southern sky, when at the highest point on the meridian, probably the most magnificent star-view in either hemisphere. This beautiful stellar view is much enhanced by the paucity of stars of large magnitude above and below these three constellations, which shine forth therefore all the brighter by the contrast. The general aspect of the remaining part of the southern sky is not remarkable, excepting the two isolated stars of the first magnitude, Canopus and Achernar, and the nebulous patches called the Magellanic clouds.

But the grandeur of the heavens south of the equator does not wholly depend on stars which never appear above the horizon of London, for it is greatly increased by well-known constellations visible in both hemispheres. For instance, the beautiful group of Orion, which adorns our evening sky in the winter months, also adds considerably to the brilliancy of that of every other part of the globe south of the equator. This constellation is even more favourably situated there than in Great Britain, on account of the greater altitude of its stars. Sirius, the brightest of all fixed celestial objects, shines more resplendently in the southern hemisphere for the same reason, being, at the Cape, when on the meridian, only eighteen degrees north of the zenith. It is therefore this combination of the great cluster of brilliant objects in Centaurus, Crux Australis, and Argo Navis, with certain others common to both hemispheres, which makes up that magnificent display which has deservedly given to the sky of the southern hemisphere that reputation for brilliancy which has been handed down from generation to generation.

We can well understand the feelings of voyagers to our southern colonies on gradually leaving behind them the last visible links which bind them to their native land; but if it were ever our fate to be in their position, we think that we should be amply compensated by the appearance, on the opposite side of the heavens, of star after star, with the configurations of which we had not been previously acquainted. The observation of the daily additions on the south horizon would be a most interesting and instructive amusement during the voyage from Europe to the Cape, as the vessel pursued its silent course towards its destination. Suppose, for example, that our voyage commenced in February, that we have taken our farewell of the Lizard Lights, that we have passed over the stormy Bay of Biscay, and that we have arrived in the latitude of Madeira, whose balmy air attracts so many to its shores. With what eagerness we should then look out for the chief star in Argo Navis, Canopus, skirting the south horizon in the early evening; and as the vessel gradually approached the equator, our progress southward would be clearly marked by the daily

LOOKING NORTH, FEBRUARY 15. (*Port Elizabeth, Algoa Bay.*)

INDEX-MAP.

LOOKING SOUTH, FEBRUARY 15. *(Table Bay, Cape of Good Hope.)*

INDEX-MAP.

increasing altitude of this star. In about latitude 20° N., several other bright objects in Argo Navis can also be identified a few degrees above the horizon. But with what an intense desire we should watch for the Southern Cross, whose stellar fame has been chronicled by navigators since the days when Vasco da Gama and his companions beheld its glittering gems from the deck of the first vessel which reached India round the Cape of Good Hope. About the end of March, at midnight, the principal star in the Cross is exactly on the horizon due south in latitude 27° 39′ N. It would not, however, be visible then; but a few degrees nearer the equator, the group may be discerned for a short time each night. By the time we have arrived on the equinoctial line, the chief object in this constellation is 27° 39′ above the horizon when on the meridian, and the stars composing the group are now very conspicuous. This part of the sky appears more brilliant as the Ship, Centaur, and Southern Cross, approach the zenith. At the Cape of Good Hope, for the latitude of which the diagrams have been prepared, these constellations, however, never reach the zenith.

The general aspect of the southern sky has, in many instances, made a deep impression on those travellers who, for the love of geographical discovery, have spent months, or even years, in unknown tropical regions. Many have given utterance to the delight they have experienced, when far away from friends, in having "sweet converse with the stars." But they have looked at the stars in a practical as well as in a poetical spirit; for travellers not only fix their latitude and longitude by their sextant observations, but the whole heavens also serve as a universal timepiece, being really and truly a great star-clock. "How often," remarks Humboldt, "have we heard our guides exclaim, in the savannahs of Venezuela, or in the deserts extending from Lima to Truxillo, 'Midnight is past, the Cross begins to bend!'" Captains Speke and Grant tracked their way through the unknown regions of Africa from Zanzibar to Gondokoro, by an almost nightly reference to the stars for the determination of the latitude and longitude of their position. Dr. Livingstone has done the same in his numerous African journeys. M. Du Chaillu informed the author that when full of disappointment and grief at the failure of his most cherished hopes during his last journey to Ashango Land, his chief solace was to lie prostrate for hours and contemplate the beauty of the heavens above him, looking upon the objects composing the brilliant scene as his companions in travel. In the published account of his travels he remarks that when at Máyolo, "the contemplation of the heavens afforded me a degree of enjoyment difficult to describe. When every one else had gone to sleep, I often stood alone on the prairie, with a gun by my side, watching the stars. I looked at some with fond love, for they had been my guides, and consequently my friends, in the lonely country I travelled; and it was always with a feeling of sadness that I looked at them for the last time before they disappeared below the horizon for a few months, and I always welcomed them back with a feeling of pleasure which, no doubt, those who have been in a situation similar to mine can understand. I studied also how high they twinkled, and tried to see how many bright meteors travelled through the sky, until the morning twilight came and reminded me that my work was done, by the then visible world becoming invisible. At the period of the year I spent at Máyolo, the finest constellations of the southern hemisphere were within view at the same time—the constellation of the Ship, of the Cross, of the Centaur, of the Scorpion, of the Greater Dog, and the Belt of Orion—which include the three brightest stars in the heavens, Sirius, Canopus, and Alpha Centauri."

Our diagrams of the sky of the southern hemisphere are intended to give a representation of the heavens as viewed from any station situated in 34° south latitude. We have chosen this terrestrial parallel, because it conveniently happens for our purpose that several important districts in Australia and the Cape of Good Hope lie exactly in this latitude. Sydney, the seat of Government in New South Wales, Adelaide, the capital of South Australia, and Cape Town, are on, or very near it; while Melbourne, in Victoria, and Perth, in West Australia, are situated within a few degrees. The northern coasts of New Zealand and many important cities in South America are also near this parallel of latitude, including Monte Video, Buenos Ayres, Santiago, and Valparaiso. Now our sky-views are equally available for all these places; for it matters not where an observer may be stationed, so long as he is on, or near, that imaginary line encircling our globe thirty-four degrees south of the equator. In the February diagrams the observer is supposed to be located at the Cape; first in the neighbourhood of Cape Town, looking south over Table Mountain, and secondly at sea, on the point of entering Port Elizabeth, in the view, looking north. But although we have arbitrarily selected these towns for our landscapes, it must not be considered that our remarks and diagrams are intended exclusively for the Cape of Good Hope; but, on the contrary, we repeat that they can be studied with equal advantage by all who wish to learn the configurations of the constellations, whether the student be in Australia or in any other place situated near the same degree of south latitude.

In describing the diagrams, we hope to point out, in as clear a manner as possible, the positions, both relatively and absolutely, of all the principal stars. We are certain that many who have taken an interest in the stars of the northern hemisphere, and in the descriptive notes and diagrams of the "Midnight Sky at London," will find equal interest in becoming slightly acquainted with the names and positions of some of the most conspicuous of the stars which, although never rising above the horizon of Great Britain, are the nightly companions of friends in far distant lands. We may quietly remain at home, but our imaginations can, with very little trouble and study, soon supply a tolerably accurate representation of the firmament invisible to us. Although the author has never seen many of the brilliant objects whose positions in the heavens he is about to describe, yet probably very few of the oldest inhabitants of the southern hemisphere are as perfectly acquainted with their exact locality; a knowledge he has obtained by the careful computation of the position of each star in the diagrams, small and great. If he were suddenly transported to the Cape of Good Hope, or Australia, he would be able to distinguish the different constellations at sight, as well, or nearly as well, as those with which he is acquainted in this country. During the preparation of the diagrams and descriptive notes of the sky invisible in Great Britain, he has clearly seen in imagination the glorious groupings of Argo Navis, the Centaur, the Southern Cross, the extra brilliancy of the Milky

Way, and the contrasting darkness of the south polar regions. No inhabitant of the southern half of our globe who is not an astronomer can possibly have more accurate conceptions of the general aspect of the southern sky than those the writer has experienced in England, while laying down star after star in the diagrams representing that portion of the heavens.

Let us first direct our attention to the diagram in which the south pole of the heavens is included. Passing down the meridian from the zenith, we come at once upon that splendid constellation Argo Navis, extending from Canopus in the west, nearly to the Southern Cross, east of the meridian. As most of the stars within forty degrees of the south pole were unknown to the ancients, those of the greatest magnitude have been catalogued by modern astronomers under the name of the constellation, with a Greek letter attached. In the days of Ptolemy and the Arabian astronomers the chief stars in Eridanus and Argo Navis were known by the names of Achernar and Canopus respectively, an appellation by which they continue to be identified to this day. Argo Navis contains two stars of the first magnitude, Canopus and Beta Argûs, five of the second, and several of the third. The reader can distinguish for himself each of these in the diagram, by comparing the constellation generally with the corresponding Greek letters in the index-map. If we now keep our eyes on the plane of the meridian, from near Beta Argûs to the south horizon, we shall pass over a most unprolific region till we fall on the constellation Toucana near the horizon. The south polar star, which is in the centre of this celestial void, is a very small object named Sigma Octantis, of about the sixth magnitude. We have inserted it in the diagram, and its position is indicated in the index-map by the letters S.P. Astronomers in the southern hemisphere are less favoured in their polar star than their European brethren. The advantages arising from having a bright star near the pole are very great, as one in that position is most favourable for the determination of some of the instrumental adjustments of the telescope as well as useful to the mariner for fixing his latitude at sea. One advantage is, that in England, Polaris can be observed at every hour of the day or night. The southern polar star, owing to its small magnitude, can, however, only be seen during the night; nor indeed can any star within some degrees of the pole be seen during the day time, although the space is occupied by the small constellations Octans, Chamæleon, Hydrus, and a few others.

East of the meridian, several brilliant stars are concentrated about midway between the zenith and horizon, among which Crux Australis, and Alpha and Beta Centauri, will be at once recognised. Several objects are contained within the boundaries of the Southern Cross; but the four which are so situated as to give it the appearance of a Latin cross are Alpha Crucis, in the lower part of the stem, nearest the pole; Gamma, in the upper part, and farthest from the pole; Beta, which marks the termination of the cross beam in the east, and Delta in a corresponding position in the west. Alpha Crucis, the brightest of the four, is of the first magnitude, Beta and Gamma are of the second, and Delta of the third. Immediately below Crux Australis are the two splendid objects in the lower part of the fore-legs of Centaurus. Beta Centauri is that next to the Southern Cross, the other being Alpha Centauri, one of the largest double stars in the heavens, and one of the nearest to our solar system. This double star has been frequently observed for the determination of its parallax, to which we shall again allude in a future chapter. The remaining stars in Centaurus, several of which are of the third magnitude, are inserted in the diagram east or left of its two principal objects and Crux Australis. Below Alpha Centauri, three moderately bright stars point out the position of Triangulum Australis, the principal star Alpha being nearest to the horizon. In the left-hand lower corner of the diagram, the stars belong to Ara and Scorpio. Very near the south horizon, east of the meridian, the chief star is Alpha Pavonis. The Milky Way is very brilliant about the Southern Cross and the Centaur.

West of the meridian line the sky is less attractive than that we have just described; but there are, however, several objects worthy of notice. Argo Navis is now on the meridian; but Canopus, its principal star, is considerably advanced towards the south-west, and is somewhat isolated from its companions in Argo. Looking due west, a part of Canis Major is inserted in the right-hand upper portion of the diagram. Below Canis Major, towards the south-west, is Columba Noachi, and near the horizon in the S.S.W., Achernar is visible. The two Magellanic clouds are in this quarter of the sky, the upper being named the Nubecula Major, and the lower the Nubecula Minor. The numerous small stars in the neighbourhood of the Magellanic clouds belong to several constellations which are only of interest to the astronomer.

Let us now briefly consider the diagram of the sky, looking north. Here the observer is assumed to be entering the harbour of Port Elizabeth, Algoa Bay, Cape Colony. All the stars inserted in this diagram are also visible at London. The principal first-class objects are Sirius, Procyon, Castor, Pollux, Alpha Hydræ, Regulus, Denebola, and Spica. Now as the relative positions of these stars on the celestial sphere always remain the same, they are consequently viewed from opposite directions in the northern and southern hemispheres. For this reason, if we go to the Cape of Good Hope, we shall find that all the constellations with which we are so familiar in England, and which are common to both hemispheres, are upside down. For example, if the reader will refer to the February diagram of the midnight sky looking south at London, he will see that the group of stars on the meridian known by the name of the Sickle, is just in a reversed position to the corresponding group in the present diagram; and so it is the same for the other constellations. West of the zenith, Sirius and other stars in Canis Major are very conspicuous. Lower down, towards the left side of the diagram, is Procyon, and lower still, the twin stars Castor and Pollux. Orion is above the horizon, but is on the point of setting. The bright isolated star above Regulus, near the meridian, is Alpha Hydræ, or Alphard. The constellations Leo and Virgo are very prominent; the names of the principal stars in them are given in the index-map. The four stars above Virgo, in the form of a quadrilateral, near the upper part of the diagram, belong to Corvus. The greater part of the space between Corvus and the meridian is occupied by Crater. A portion of Ursa Major lies between Leo and the north horizon, its position being indicated by the two close pairs of stars, Lambda and Mu Ursæ Majoris, and Iota and Kappa Ursæ Majoris, near the horizon. Cor Caroli is very low down in the north-east. Arcturus and the principal stars in Boötes are shining conspicuously in the N.N.E., but this portion of the sky is too far from the meridian to be included in this month's diagram.

The reader is reminded, in this place, that the seasons in the southern hemisphere are in reverse order to those in the northern. As the axis of the Earth is inclined about twenty-three degrees to the direction of the Earth's motion around the Sun, the north pole of the Earth is turned towards the Sun at the summer solstice, while the south pole is nearest the Sun at the winter solstice. The effect of this is that the seasons of the two hemispheres are always in opposition to each other. In Great Britain the longest day takes place on June 21st, and the shortest on December 21st; but in Australia, the Cape of Good Hope, and in the lower part of South America, the shortest day takes place on June 21st, and the longest on December 21st. Consequently, when it is summer in England, it is winter in the southern hemisphere, and *vice versâ*.

MAY.

—◆—

AT midnight, from February to May, the four chief stars in Crux Australis, the two brightest stars in Centaurus, and the most attractive portion of the Milky Way, are all in their most favourable positions for observation, being at that time on, or slightly east or west of, the upper meridian. The exact day when the Southern Cross was on the meridian at midnight was on March 27th, the group then being vertical in the direction of the south pole. In the middle of May, however, at the hour represented by our diagrams, the Cross has passed the meridian towards the west, and is commencing a downward course in the direction of the south horizon. In fact, all the stars which were so attractive at their most elevated positions in February have made considerable progress onwards in their apparent annual revolution around the south pole. It may be useful to remind the reader in this place of what we have already explained in our description of the sky of the northern hemisphere for March, that these daily, monthly, or quarterly changes which we notice in the position of the stars with respect to the zenith and horizon at any fixed hour, are due entirely to a corresponding daily, monthly, or quarterly change in the position of the Sun on the ecliptic, and not to any real variation in the positions of the stars themselves. After the expiration of a year, when the Sun has made an apparent complete revolution on the ecliptic, the Earth will be in the same part of its orbit as on the corresponding day of the preceding year, when, as the relative positions of the Sun and Earth will be the same, similar configurations of the stars will be visible. We have also mentioned previously that this monthly or seasonal change is quite independent of the hourly change produced by the rotation of our globe on its axis. Every one knows, or ought to know, that all the stars make apparently a complete revolution around the Earth in twenty-four hours. This is very evident to the youngest reader of these pages. For example, we see the stars which are not circumpolar rise in the eastern horizon; then gradually ascend till they reach their highest point on the meridian; then as gradually descend towards the western horizon, where they disappear from view for several hours. While invisible to us, they are passing in like manner across the sky visible from the opposite side of the globe, and in due time they will reappear on our eastern horizon to go through their regular course again, without the remotest chance of erring. As the stars are fixed, all these apparent movements are solely produced by the turning of the Earth.

W

E

LOOKING NORTH, MAY 15.

β Corvi
γ.
δ
Antares
δ.
β Scorpii
η Ophiuchi
λ Aquilæ
Spica
α Libræ
β
η
γ Virginis
β Libræ
ε Ophiuchi
δ
β Ophiuchi
ε
α Serpentis
Ras Alague
ζ Aquilæ
δ
α Herculis
ε
γ.
Arcturus
β Herculis
ε Boötis
Alphecca
ζ
δ Herculis
β Boötis
η
π̈
β Lyræ
χ
Cor Caroli
Vega
Alkaid

INDEX-MAP.

110

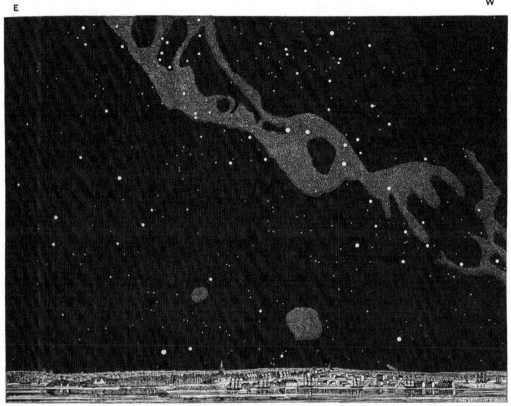

LOOKING SOUTH, MAY 15.   *(Sydney, New South Wales.)*

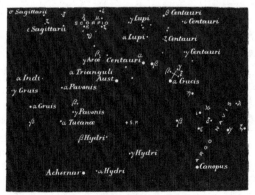

INDEX-MAP

We are, indeed, travelling in this manner, independently of our orbital motion, with a most astonishing rapidity; for we, and everything on the surface of the Earth, are spinning round like a top at the rate of 1,040 miles per hour, and yet to our ordinary feelings we are perfectly unconscious of it, so uniform is the motion.

As an illustration of the daily movements of a south circumpolar star, one which is situated so near the south pole as to be always above the horizon during the twenty-four hours, we will select the group of four stars composing the Southern Cross, so as to exhibit the position of that constellation, with respect to the zenith, at different hours on May 15th and 16th. It will be observed that, in the course of the day, the constellation will have made a complete circuit around the south pole. The four principal stars by which it is generally distinguished are between the zenith and the south pole, and on the upper meridian exactly at 8.45 P.M. on May 15. At 2.44 A.M. on May 16 the Earth will have turned on its axis through one quarter of its revolution, the stars will therefore apparently have passed over one quadrant, or the fourth part of the circuit, being at that time due west of the south pole. At 8.43 A.M. they have performed one-half of their circuit, being now near the horizon, on the lower meridian. At 2.42 P.M. they are due east of the south pole, while the complete revolution is made at 8.41 P.M. At the hour of midnight this constellation is in the four positions, north, west, south, and east of the south pole, at the end of March, June, September, and December respectively.

We will now devote our attention to an explanation of the two sky-views. As usual, we confine our remarks in the first place to the diagram of the sky looking south, over Sydney, New South Wales. Taking that portion of the heavens west of the meridian, or on the right-hand side of the diagram, it will be at once perceived that the constellations to whose brilliancy the southern sky owes its pre-eminence are all situated in that direction. Commencing at the zenith, and passing onwards in order towards the south-west horizon, we shall include nearly every principal star of note. Before we reach the Southern Cross, the space is principally occupied by the constellations Centaurus and Lupus, the stars of which are somewhat mingled. This group consists of several stars of the third magnitude, about equally divided in number between the two constellations. The two principal stars in Centaurus are both easily recognised above Crux Australis, Alpha Centauri, the celebrated double star, being that nearer the meridian, while Beta Centauri is between it and Beta Crucis, the most easterly of the four principal stars in the Cross. The position of the Southern Cross itself is always well marked in the heavens. If the reader will refer to the February diagram of the southern hemisphere, he will perceive there that it is situated at a considerable elevation, but on the left side of the meridian, a line drawn between the upper and the lower stars being inclined away from the meridian towards the north-east. In the May diagram the inclination of the Cross, which is now on the right side of the meridian, points in a north-westerly direction. When on the upper meridian, between the zenith and south pole, the Cross is, as we have previously remarked, in a perpendicular position, the lowest star being the brightest of the group. When on the lower meridian, the position of the stars appears reversed, the principal star then being the highest. Captain Basil Hall, during a cruise in the southern ocean in a latitude very near to that for which our diagrams are constructed, refers to the varying position of the stars of the Southern Cross as seen from his ship at sea. " I have observed it," he remarks, "in every stage, from its triumphant erect position, between sixty and seventy degrees above the horizon, to that of complete inversion, with the top beneath, and almost touching the water. This position, by the way, always reminded me of the death of St. Peter, who is said to have deemed it too great an honour to be crucified with his head upwards. In short, I defy the stupidest mortal that ever lived to watch these changes in the aspect of this splendid constellation, and not to be in some degree struck by them."

Between the Southern Cross and the south-western horizon the sky is occupied entirely with the constellation Argo Navis. Of this remarkable collection of stars, Canopus, near the horizon, and Beta Argûs, the nearest of the group to the meridian, are the principal. A reference to the index-map will point out to the reader the relative positions, as well as the names, of most of the others. The brilliancy of the south-western sky is much increased by the addition of the most luminous portion of the Milky Way, which passes through Centaurus and the Southern Cross. The Nubecula Major, or principal Magellanic cloud, is near the horizon, slightly west of the meridian. Excepting the three stars in Triangulum Australis, there are no others on, or very near, the meridian of sufficient magnitude to attract attention. Triangulum Australis is south-east of Centaurus. The minute south polar star, Sigma Octantis, is almost in the same position as in February.

East of the meridian, or the left half of the diagram, the appearance of the sky shows a very marked contrast to that on the opposite side of the meridian, although there are several stars of the first and second magnitudes scattered here and there. Looking due east from the zenith, or near the upper part of the diagram, a considerable number of stars of the third magnitude mark the position of the southernmost portion of Scorpio. Antares and Beta Scorpii, the two principal stars in this constellation, are north of the zenith ; they are consequently to be found in the diagram, looking north. Due east of the group in Scorpio, several stars in Sagittarius may be noticed. Below Scorpio in the diagram, and between it and Triangulum Australis, the eye falls on Ara, the altar, containing several tolerably bright stars, some of which are of the third magnitude. The principal constellations in the south-eastern sky, between Ara and the horizon, are Pavo, Toucana, Grus, and Indus. Their relative positions may be gathered from the names of their principal stars inserted in the index-map. Alpha Pavonis and Alpha Gruis are of the second magnitude. Hydrus is below the south pole, and near the horizon a part of Eridanus is situated. Its principal star, Achernar, is now several degrees east of the lower meridian, and very near the horizon. This star and Canopus in Argo Navis are the only two which never rise above the horizon of Europe, whose names have been derived from the ancients. The Nubecula Minor is in this part of the sky between Achernar and the south pole. In addition to the principal constellations which have come under our notice, there are numerous others, but scarcely one contains a single star above the fourth magnitude.

Several of these surround the pole, thus making that district of the heavens so celebrated for its comparative darkness, which appears all the more obscure by its contrast with the brilliancy of other portions of the sky.

We will now say a few words in illustration of the diagram looking north. Most, if not all, of the stars which will come under our notice are common to both hemispheres, and are consequently familiar by name after the perusal of the chapters on "The Midnight Sky at London." The principal constellations included in the diagram are Hercules, Boötes, Corona Borealis, Lyra, Virgo, Serpens, Ophiuchus, Libra, Corvus, Scorpio, Aquila, and the southern part of Ursa Major. Serpens and Corona Borealis are on the meridian, the former being midway between the zenith and the north horizon, the position of the head being indicated by a group of three stars, the central one of which is Alpha Serpentis. Alphecca, the chief star in the Northern Crown, is exactly between Alpha Serpentis and the horizon. East of the meridian, the principal stars are Ras Alague, Alpha Herculis, Vega, Antares, and Beta Scorpii. All these, together with those on the opposite side of the meridian, can be identified by comparing the stars in the large diagram with those inserted in the index-map. West of the meridian, Spica, Alpha and Beta Libræ, Arcturus, and several average-sized stars in Virgo and Boötes, may be specially mentioned. Alkaid, or Eta Ursæ Majoris, the star which points out the position of the tip of the tail of the Great Bear, is just visible above the horizon in the N.N.W. The relative positions of the constellations being the same as those in the corresponding map of the stars visible in the northern hemisphere in May, we must refer the reader to the detailed descriptions there given for special remarks on the different constellations. But it will be necessary for us to state that the stars which are in the zenith of England are near the north horizon at the Cape of Good Hope, Australia, Buenos Ayres, etc., and those which are in the zenith of these places are visible in the south horizon of England.

In the southern hemisphere the apparent different positions of the Sun, Moon, and planets, from those observed in northern latitudes, is sure to attract the attention of the most casual observer. In this country, we are all accustomed to look for these objects, when on the meridian, with the face directed due south of the zenith. At the Cape of Good Hope, or in Australia, however, the position of the observer is exactly reversed. When, therefore, the principal planets are at their greatest southern declination, and when their right ascension is nearly the same as that of Argo Navis or Crux Australis, the combined brilliancy produced by the planets north, and the stars north and south of the zenith, is wonderfully increased. Perhaps one of the most brilliant stellar spectacles in Great Britain is the appearance of the eastern sky at midnight in the early part of November. We have always had this impression on our mind when looking in that direction of the heavens year after year, for the appearance of the meteors which form part of the ring of the memorable stream observed so universally in England in 1866, and in North America in 1867 and 1868. At these times we have had before us the glorious constellation of Orion, the thickly-studded Taurus, the twin stars Castor and Pollux, the seven stars in Ursa Major, and several others of the first class, including Sirius, Capella, Procyon, and

Regulus. All these objects combined constitute a scene of more than usual stellar beauty. But notwithstanding this magnificent exhibition of our principal stars, we believe that a spectator placed in a locality several degrees south of the equator, where he can enjoy the mild radiance and the external tranquillity of the tropical nights, and where he has the benefit of not only most of the brilliant stars visible to us, but also others, equally brilliant, invisible in northern latitudes, never fails to award the palm to the equatorial sky. For the brightly-beaming constellation Argo Navis, the Milky Way between Scorpio and Centaurus, with its contrasting bright and dark patches, the ever-popular Southern Cross, the two neighbouring first-class stars in the legs of the Centaur, and, indeed, we may add, the picturesque beauty of the whole expanse of the heavens north and south of the zenith, are materials amply sufficient to leave upon the mind of the most indifferent observer, such deep impressions of the majesty of creation as ought not to be easily effaced.

AUGUST.

HE stars which have been so conspicuous in the upper portions of the diagrams for February and May, looking south, are now all near the horizon, some in Argo Navis being below that limit. The sky at midnight in the winter months of Australia, the Cape of Good Hope, and other parts of the southern hemisphere in the same latitude, is therefore not of that brilliant character which we have described in former months. Most of the principal stars inserted in preceding diagrams are, however, visible, but much of their attractiveness is lost by reason of their proximity to the horizon. If at this season of the year the general aspect of the heavens to the naked eye is less striking than at midnight in February, the superior definition of the stars, owing to the comparatively lower temperature of the air, enables the astronomical observer to carry on his delicate observations with far greater precision than when the images of the stars are in a tremulous state, caused by the excessive heat and dryness of the sandy plains. When Sir John Herschel was at the Cape of Good Hope, in the years 1834 to 1838, he frequently noticed that in the cooler months, from May to October inclusive, and more especially in June and July, that the finest opportunities for delicate astronomical observation occurred. He remarks that the state of the atmosphere in these months, as regards definition, was habitually good, and imperfect vision was rather the exception than the rule. The best nights, when the stars were most steady, always occurred after the heavy rains, which generally fell at this season, had ceased for a day or two, when "the tranquillity of the images and sharpness of vision was such that hardly any limit was set to magnifying power but what the aberrations of the specula necessitate."

In our examination of the diagram looking south, in detail, beginning at the zenith, which is now occupied by the constellation Piscis Australis, we will follow at first the course of the meridian. In doing this, we shall pass through Grus, Toucana, Pavo, Hydrus, and Argo Navis, besides several small constellations. A star of the third magnitude near the zenith and the meridian is Gamma Gruis; that immediately below it is Alpha Gruis, while the star a little to the left of Alpha is Beta, of the same constellation. The principal star in Toucana is below Grus, slightly east of the meridian, and those slightly west of the meridian are the principal objects in Pavo. The small star, Sigma Octantis, pointing out the position of the south pole, is midway between Pavo and Argo Navis, the

117

latter constellation being partly above and partly below the south horizon. The principal stars, near the meridian, visible at this time in Argo Navis, are Beta Argûs, and several others, the names of which can be identified by reference to the index-map.

West of the meridian, several very prominent stars are visible. This portion of the sky now includes, among others, the following principal constellations:—Pavo, Indus, Scorpio, Ara, Triangulum Australis, Centaurus, and Crux Australis. The brightest star in this direction of the heavens nearest the zenith is Alpha Pavonis. In a due westerly direction, at the right-hand upper portion of the diagram, most of the small stars belong to Sagittarius. Directly below these, a moderately-bright group consists of a portion of Scorpio, whose principal stars, Antares and Beta Scorpii, although not included within the limit of the diagram, are still visible in the heavens, approaching the horizon in W.S.W. Near the Scorpio group, but a little lower, Ara is situated, its principal star, Alpha Aræ, being that nearest the group. Beta and Gamma Aræ, of nearly the same magnitude, are the two stars very near each other, Beta being the more northerly of the two. Below Ara, the three principal stars in Triangulum Australis are easily identified by their relative positions. The south-west corner of the diagram is occupied by several conspicuous objects in Centaurus and Lupus, two adjoining constellations. Alpha and Beta Centauri are now the most easterly of the stars in Centaurus. The Southern Cross is near the horizon in S.S.W., embosomed in the Milky Way, which at this time passes across the western sky.

East of the meridian, the principal constellations are Phœnix, Grus, Eridanus, Hydrus, a considerable part of Argo Navis, and the two Magellanic clouds. South of Gamma Gruis, near the zenith, Alpha Gruis, of the second magnitude, is the first object to attract attention. Due east of Alpha Gruis is the second star of that constellation; and still farther east, or to the left, is Alpha Phœnicis, also of the second magnitude. Immediately below Alpha Phœnicis is the brilliant Achernar, the chief star in Eridanus; and in the same direction, near the horizon, the equally lustrous Canopus, the leading star in Argo Navis, shines above all others in that quarter of the sky. The Nubecula Minor, or smaller Magellanic cloud, may be noticed between Achernar and Beta Hydri, but nearer to the latter star. The Nubecula Major is below Gamma Hydri in the direction of Canopus. The intermediate small stars between Hydrus and Eridanus belong to several small constellations.

In the diagram looking north, the observer is supposed to be at sea, in latitude 34° south, with his face turned in the direction of the equator. Passing the eye down gradually from the zenith to the north horizon on the meridian, several stars in the following constellations may be identified. From Piscis Australis in the zenith, we pass over the eastern portion of Capricornus, Aquarius, Pegasus, and Cygnus. No very bright stars are, however, near the meridian, excepting perhaps Epsilon Pegasi, which is just midway between the zenith and horizon. East of the zenith, and about twelve degrees from it, is Fomalhaut, a star of the first magnitude. In the same direction, and near the right-hand upper corner of the diagram, Beta Ceti and a few other stars in Cetus may be distinguished. The north-eastern portion of the sky is chiefly occupied by Aquarius,

LOOKING NORTH, AUGUST 15. (*Sydney North Head.*)

INDEX-MAP.

E   W

LOOKING SOUTH, AUGUST 15. *(Table Mountain, under the South-east Cloud.)*

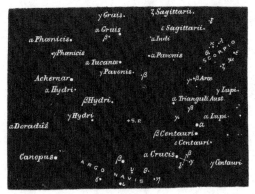

INDEX-MAP.

Pegasus, Pisces, and Andromeda. Pegasus extends from Epsilon Pegasi on the meridian as far as Gamma Pegasi, the south-western corner of the square of Pegasus. The position of this stellar quadrilateral is very easily recognised. The north-eastern star in the square is Alpherat, the two to the right of it being Delta and Beta Andromedæ respectively. The small stars contained in Pisces are east and south-east of Pegasus.

West of the meridian north of the zenith, the principal constellations are Capricornus, Aquila, Delphinus, Vulpecula, Lyra, Ophiuchus, and portions of Serpens, Cygnus, Hercules, and Sagittarius. The most prominent stars are the group of three in Aquila, Vega, Alpha Cygni, Albiero, and Ras Alague, or Alpha Ophiuchi. The last mentioned is, however, just without the limit of the diagram. The two chief stars in Capricornus, owing to their great elevation, are sure to attract the eye. Their position is always pointed out in the heavens by the three stars in Aquila. A line drawn through these three stars leads in one direction nearly to Alpha and Beta Capricorni, while another drawn in the opposite direction indicates the position of Vega and the other stars in Lyra. The small group consisting of the largest stars in Delphinus may be recognised between Aquila and the meridian. Taken as a whole, a glance shows that the general aspect of the midnight sky of the southern hemisphere, both north and south of the zenith, is much inferior in the months of July, August, and September, to that exhibited in the preceding and following months of the year. In August the brilliant stars in Argo Navis, Crux Australis, and Centaurus are in their most elevated positions during daylight, when they are invisible to the unassisted eye. In the early evening and morning hours, these constellations are, however, very conspicuous objects in the south-west or south-east.

NOVEMBER.

—◆—

THE eastern portion of the sky above the horizon at midnight in the middle of November contains the greater number of the most brilliant constellations, visible both in the northern and southern hemispheres. What a magnificent stellar spectacle is spread out before an observer who, with his face directed towards the east, is at this moment located in any part of the Cape of Good Hope, Australia, Tasmania, or New Zealand! In South America also, in the neighbourhoods of Buenos Ayres, Monte Video, Santiago, and Valparaiso, the same brilliancy of the eastern sky can be seen at the local midnight of each place. Before, however, entering into our usual detailed explanation of the diagrams, let us for an instant give in imagination a passing glance at the different constellations and stars of the first magnitude, which collectively make up this wonderful display. As a stellar view it far excels that to be seen at the same time in our eastern northern sky, although many of the principal constellations are common to the two hemispheres. From Centaurus, near the horizon in the S.S.E., to Leo, near the horizon in the N.N.E., one continuous succession of splendid grouping of large stars may be noticed, without many intermediate dark spaces. First, we have the two chief stars in the Centaur, then the popular Southern Cross, afterwards the extensive and prolific constellation Argo Navis; these are followed by our own Canis Major, Orion, Canis Minor, Taurus, Gemini, Leo, and Auriga. Included within these, we have at this moment exhibited to our view the following well-known stars of the first magnitude:—Alpha and Beta Centauri, Alpha Crucis, Canopus, Sirius, Rigel, Betelgeuse, Aldebaran, Capella, Castor, Pollux, Procyon, and Regulus. To these may be added many others only one grade inferior in magnitude, and also the most brilliant portion of the Milky Way, which is, however, very near the horizon in the S.S.E. The midnight sky of December and January is quite equal, if not superior, to that of November, as most of the principal stars we have just mentioned are in more elevated positions in the two former months.

The first diagram to which we will give our attention is that in which the reader is supposed to be looking south over Buenos Ayres. Our friends in Australia, or at the Cape of Good Hope, must not, however, forget that our explanations are intended as much for them as for those in the Argentine Republic, or on the northern and southern shores of the Rio de la Plata. In the zenith there are no stars of large magnitude, that spot being

124

W E

LOOKING NORTH, NOVEMBER 15. (*Monte Video.*)

INDEX-MAP.

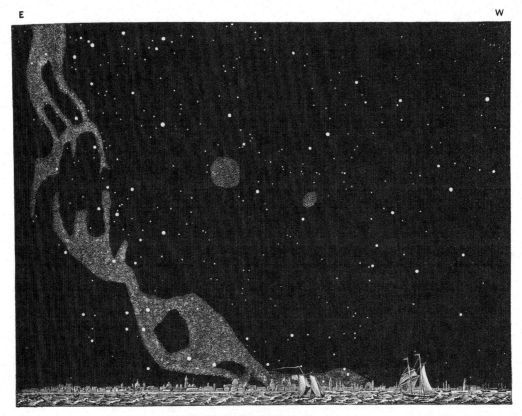

LOOKING SOUTH, NOVEMBER 15. *(Buenos Ayres.)*

INDEX-MAP.

occupied by a very unprolific part of Eridanus. Those on the meridian are also very inferior in magnitude. Gamma Hydri, between the two Magellanic clouds, is the brightest star near that imaginary line, until within a short distance of the horizon, where the meridian passes through the small constellation Triangulum Australis, distinguished by three moderately bright stars in the form of a triangle to the right of Alpha Centauri. If we now take a survey of the south-western portion of the sky, or the right-hand side of the diagram, we shall perceive a fair sprinkling of bright objects, although they are not contained in any remarkable groups. The three bright objects in the upper part of the diagram are Achernar, the nearest of the three to the meridian, Alpha Phœnicis, and Fomalhaut. Achernar is in the southernmost part of Eridanus, a constellation of great extent, reaching as far north as the celestial equator. Fomalhaut is easily distinguished in the latitude of London, but it never rises there more than about eight degrees above the horizon. In latitudes south of the equator, this star is, however, a very prominent object. Below Fomalhaut are the constellations Grus, Toucana, Pavo, and Indus. The Greek letters, by which their principal stars can be identified, are inserted in the index-map. Near the horizon, south-west of Triangulum Australis, is a part of Ara.

East of the meridian, or the left-hand side of the diagram, contains Columba Noachi, Doradus, Argo Navis, Crux Australis, and a small part of Centaurus. The greater portion of Centaurus is now below the horizon. Due east of the zenith, the first large star is Alpha Columbæ; the important group to the left of Alpha Columbæ is in the southern part of Canis Major, whose principal star, Sirius, is inserted in the upper diagram. The extensive group between Canis Major and Crux Australis forms the constellation Argo Navis. Canopus, its principal star, is the nearest to the upper meridian. The most remarkable object, however, in Argo Navis is distinguished by the name of Eta Argûs, near the Southern Cross. It is not only a most extraordinary variable star, but it is situated in the centre of the great nebula in Argo, with which it has been supposed to have some connection. We shall refer more in detail to these two objects in a subsequent chapter. The two nebulous Magellanic clouds are now in their most favourable positions, the larger being slightly east of the meridian, and the smaller about the same distance on the opposite side of the meridian. The brighter part of the Milky Way, near the Southern Cross, is not very favourably situated, although it may be recognised without much difficulty when the sky is very clear. As very many small southern constellations contain no bright star, it is useless to point out their positions in the diagram, or even to give their names in this place.

In the diagram, in which the observer is supposed to be looking north over Monte Video, we can recognise at sight nearly all the winter stars of the northern hemisphere, with this important difference, that they are all seen in the reverse position to what we are accustomed to observe them in England. Let us now briefly glance at the principal stars and constellations, taking the right-hand side of the diagram first. The two bright objects in the upper part are Sirius and Beta Canis Majoris. Orion is so prominent that it is scarcely necessary to indicate its position, except perhaps to remark that the left-hand upper star of the quadrilateral is Rigel, and that the right-hand lower star is Betelgeuse. Procyon, in Canis Minor, is at the extreme limit of the diagram. Taurus and Gemini are below Orion; and below Taurus, a part of Auriga is above the horizon, with its two principal stars, Capella and Beta Aurigæ. Low down in the east Leo has just risen, and several of its stars, including those in the Sickle, can be seen in the heavens; but they are outside the limit of the diagram. The Pleiades and a few ordinary stars in Eridanus and Perseus are now on the meridian. On the left side of the meridian, Cetus, Aries, Andromeda, and Pegasus are the chief constellations. Although several important stars are now in the north-western sky, yet the general aspect of this division of the heavens looks quiet in comparison with the brilliancy of that east of the meridian.

THE CONSTELLATIONS.

"ON the Earth's orbit see the various signs
   Mark where the Sun, our year completing, shines;
   First the bright Ram his languid ray improves;
   Next glaring wat'ry through the Bull he moves;
   The loving Twins admit his genial ray;
   Now burning through the Crab he takes his way;
   The Lion flaming bears the solar power;
   The Virgin faints beneath the sultry shower;
   Now the just Balance weighs his equal force,
   The slimy Serpent swelters in its course;
   The fabled Archer clouds his languid face;
   The Goat, with tempests, urges on his race;
   Now in the water his faint beams appear,
   And the cold Fishes end the circling year."

CHATTERTON.

THE SIGNS OF THE ZODIAC.

HE order of the signs of the zodiac in the heavens is as follows:—Aries, Taurus, Gemini, Cancer, Leo, Virgo, Libra, Scorpio, Sagittarius, Capricornus, Aquarius, and Pisces. The zodiac is a broad zone of the heavens within the area of which all the apparent motions of the Sun, Moon, and major planets, are included. The breadth of this zone is about seventeen degrees, or between eight and nine degrees on each side of the ecliptic, or the apparent path of the Sun in the heavens. The ecliptic is inclined to the equinoctial at an angle of 23° 28′; they consequently intersect each other twice during the year, first at the vernal equinox, and, secondly, at the autumnal equinox. The length of the zodiac is 360°, which was divided by the ancients into twelve equal portions, or signs, of thirty degrees, each sign receiving a symbolic name, by which it is distinguished to this day. Certain influences on mankind, either for good or evil, were attributed, from a very early age, to the separate signs, depending upon the position of the Sun in the ecliptic. This superstitious belief has even continued to some extent down to modern times. In addition to the twelve zodiacal signs, the visible heavens were divided by the ancients into thirty-six other asterisms, all of which exist at present under the original nomenclature. They are all adopted by Ptolemy in his *Almagest*, which includes a catalogue of stars depending in a great measure on the observations of Hipparchus, the father of observational astronomy.

ARIES.

Aries, the Ram, is the first in order in the zodiac of Hipparchus, and is consequently one of the forty-eight ancient constellations. It has the small asterisms of Triangulum and Musca Borealis on the north, Taurus on the east, Cetus on the south, and Pisces on the west. The position of Aries in the heavens can be easily found, as it contains several average-sized stars, two of which are of the second magnitude, named Alpha and Beta Arietis. These are both situated in the head of the Ram, nearly midway between Alpherat in the square of Pegasus and the Pleiades. Aries, and the neighbouring small constellation Triangulum, contain several double and triple stars of a very interesting character. Although this constellation is the first, or leading sign, in the ancient zodiac, yet by reason of the retrograde motion of the equinoctial points, termed the precession of the equinoxes, the Sun, which formerly entered Aries on the 20th of March, reaches that point much later in the year. Owing to this, the point of the heavens intersected by the celestial equator and ecliptic, technically still called "the first point of Aries," is really far advanced into Pisces, which is strictly the first sign in the zodiac of the present age. However, for the sake of uniformity, it is not likely that astronomers would wish to alter the name of the zero-point from which all right ascensions are measured, or to disturb the order of the zodiacal signs, as settled by the ancients, notwithstanding that Pisces is the habitation of the Sun at the beginning of his annual course.

In Ptolemy's catalogue, which includes the stars observed by Hipparchus, eighteen stars in Aries are inserted. The atlas of Bode contains one hundred and forty-eight, including those visible through a small telescope. Viewed with some of the large and space-penetrating instruments of the present time, this number may be increased to thousands. The same remark applies to each constellation.

TAURUS.

Taurus, the Bull, the second of the zodiacal signs, is bounded by Perseus and Auriga on the north, Gemini on the east, Orion and Eridanus on the south, and Aries on the west. It is composed of a great number of stars, no fewer than four hundred being contained in the atlas of Bode; the majority are, however, of small magnitude, visible only by the aid of an ordinary astronomical telescope. Aldebaran is the principal star, situated in the midst of a group of smaller objects called the Hyades.

A well-known cluster, and perhaps the most popular group of stars in any part of the sky, is that known by the name of the Pleiades, situated in the shoulder of Taurus. This cluster has been mentioned in poetry as far back as Hesiod, who alludes to them as the Seven Virgins. In the ancient manuscript of Cicero's *Aratus*, preserved in the British Museum, the stars are named Merope, Alcyone, Celieno, Electra, Maia, Asterope, and Taygeta. Though they have been named the " seven stars," yet to ordinary eyes six only are visible. On brilliant moonless nights, however, not only have the ancient seven been perceived, but several more. An observer, the discoverer of the new star of 1604, saw occasionally fourteen without any glasses, a feat which has not been repeated by any other person. A member of the family of the present Astronomer Royal has habitually seen seven stars, and on very rare occasions, when the sky has been unusually clear, twelve have been distinguished. On February 15th, 1863, a map of the stars was drawn as viewed by the unarmed eye, without knowing at the time the actual relative positions of the different stars. On comparing this map with one constructed from the telescopic measures of M. Bessel, there was no difficulty in identifying the twelve stars as amongst those which Bessel had named by certain arbitrary numbers. A tolerably good telescope will exhibit about a hundred stars in the Pleiades.

A very strong interest in this remarkable group has been excited in many minds by the scriptural mention of them in the book of Job, conjointly with a few other stars, while the patriarch is being convinced of his ignorance and imbecility—

" Canst thou bind the sweet influences of Pleiades, or loose the bands of Orion ?
    Canst thou bring forth Mazzaroth in his season ? or canst thou guide Arcturus with his sons ?
    Knowest thou the ordinances of heaven ? canst thou set the dominion thereof in the earth ?"

The remarks of Admiral Smyth on these words will, we are sure, be interesting :—
" Now this splendid passage, I am assured, is more correctly rendered thus :—

'Canst thou shut up the delightful teemings of Chimah ?
    Or the contractions of Chesil canst thou open ?
    Canst thou draw forth Mazzaroth in his season ?
    Or Aish and his sons canst thou guide ?'

In this very early description of the cardinal constellations, Chimah denotes Taurus, with the Pleiades ; Chesil is Scorpio ; Mazzaroth is Sirius, in the ' chambers of the south ;' and Aish the Greater Bear, the Hebrew word signifying a *bier*, which was shaped by the four well-known bright stars, while the three forming the tail were considered as the children attending a funeral. St. Augustin, in his annotations on the above passage, assures us that, under the Pleiades and Orion, God comprehends all the rest of the stars, by a figure of speech, putting a part for the whole ; and the argument is that the all-powerful Deity regulates the seasons, and no mortal can intermeddle with them or presume to scan the ordinances of heaven."

GEMINI.

The constellation Gemini, the Twins, is the third member of the zodiac, having Taurus on the west, and Cancer on the east. Its principal stars may be identified in the heavens about midway between Orion and Ursa Major. On most globes and celestial maps, the Twins are generally represented as two youths, or children, the idea of which first originated with the ancient Greeks. The Orientals, however, occasionally adopted two kids, and the Arabians two peacocks. To the working astronomer, however, these imaginary forms go for nothing : he only sees in Gemini numerous interesting objects, including several double stars, clusters, and nebulæ, with two splendid stars, Castor and Pollux. Castor is one of the most interesting double stars in the heavens. It will be remembered that the ship which conveyed St. Paul from Melita to Puteoli was named after these stars, by which it appears probable that they were esteemed by mariners in that age as propitious.

Castor, which is the more northerly of the two, consists of two white stars of nearly equal magnitude, belonging to a common system. It is what is termed a binary star. When Sir William Herschel observed it in 1778 the position angle was 302 degrees, and when it was observed in 1821 by Sir John Herschel and Sir James South, the position of the secondary star had changed its quadrant to an angle of 267 degrees. On examining the observations made by Dr. Bradley and other astronomers of the last century, and comparing them with those of more recent observers, the period of the orbit of Castor, or the time occupied in the revolution of one star around the other, has been found to be a little less than a thousand years. With a telescope of moderate power the duplicity of

135

Castor can be easily seen; to the naked eye, however, it shines as one bright star. Pollux, the brother of Castor, is a star with an orange tinge, and is between the first and second magnitudes. It has been suspected to have been variable in its lustre at different epochs: for example, Ptolemy, Tycho, and others have classified it among those of the second magnitude; some have recorded it of the third; while Dr. Bradley speaks of it as of the first. Probably these variations are only different methods of estimation. Castor and Pollux, with the other principal stars in Gemini, are very conspicuous, as they shine in a district of the heavens free from other bright objects; the greater number of the stars in the neighbouring constellations, Cancer and Lynx, not exceeding the fourth magnitude.

CANCER.

Though Cancer, the Crab, is one of the most insignificant of the signs of the zodiac, with respect to bright stars, nevertheless it contains several very interesting objects, visible only through good telescopes. Among these is Zeta Cancri, a fine triple star near the hind claws of the Crab. Two of the stars are very close, and afford an excellent test of the power and goodness of the object-glass of an astronomical telescope, as these stars require one of the best quality to separate them. The third star is at no great distance from the other two. It has been found that the two close stars revolve around each other in about sixty years, while the outer one takes five hundred years to perform its revolution around the others. It is thus evident that these objects are not simply in juxtaposition optically, but that they belong to one system. A small nebulous-looking object in the Crab's body, visible to the naked eye on very brilliant nights, is known by the name of the Præsepe, or the Beehive. This remarkable cluster, when viewed through a telescope of low power, is resolved into an aggregation of small stars. It has been specially noticed by the ancients, particularly by the Greek philosophers Theophrastus and Aratus, who have told us "that its dimness and disappearance during the progressive condensation of the atmosphere were regarded as the first sign of approaching rain." It was formerly supposed that this group consisted only of three nebulous stars, which emitted that peculiar light alluded to by old philosophers. The Præsepe is, however, rather scanty of stars if we compare it with many other clusters; but soon after the invention of the telescope Galileo counted thirty-six small stars in the group. Cancer contains several good double stars, which are always interesting objects to the astronomer, though, from their small magnitude, some of them are scarcely visible to the naked eye.

The Sun enters Cancer on June 21st; this constellation is consequently the first of the summer signs of the zodiac. The Sun at that time is at its greatest north declination, the north pole of the Earth being turned towards it at its greatest inclination; we have therefore the longest day in the northern hemisphere. In all countries south of the equator, however, the shortest day will take place at this time, the south pole of the Earth being turned away from the Sun at midsummer.

LEO.

Leo, the Lion, the fifth sign in the order of the zodiac, is one of the principal constellations which adorn the midnight sky of spring. It is bounded, generally, on the north by Leo Minor, on the west by Cancer, on the south by Sextans, and on the east by Virgo. The principal star, Regulus, is also designated Cor Leonis, or the Lion's Heart. This star was considered by the ancients as truly royal. By the Arabs it was denominated a "fiery trigon," or lion's heart, as well as a kingly star. Wyllyam Salysbury, writing in 1552, tells us that "the Lyon's herte is called of some men the Royall Starre, for they that are borne under it are thought to have a royall nativitie." This royal star is very prominent in the spring diagrams, by the help of which there can be but little difficulty in finding it in the heavens.

Regulus, Gamma Leonis, and a few other stars of smaller magnitude, make a group whose form gives a very fair representation of a Sickle. It is from a point near these stars that a great majority of the meteors observed during the memorable star-showers of November, 1866, 1867, and 1868, were found to radiate.

The Sickle is not the only attractive group of stars in Leo. Denebola, in the extremity of the Lion's tail, and Delta Leonis, form, with Regulus and Gamma Leonis, a large trapezium. A mere glance in the direction of Leo cannot fail to identify each of these stars; but, if not, then this trapezium can be readily found by reference to the universally known pointers in the Great Bear, which serve to indicate the position of Polaris in one direction, as we have before remarked, while the same line produced in the opposite direction will pass through Leo. There are a great number of interesting telescopic objects in this constellation, visible with the aid of ordinary telescopes, especially double and variable stars. One of the latter, R Leonis, is remarkable for its blood-red appearance, which is very striking to the eye when viewed for the first time through a good telescope.

Leo contains about a hundred stars visible to the unassisted eye.

VIRGO.

The constellation Virgo, the Virgin, which occupies a considerable portion of the sky west of the meridian at midnight in April, and east of the meridian in the earlier hours of the evening in that month, is the sixth sign of the zodiac. It was popularly considered in former times as the sign belonging to the harvest season, because when the Sun enters it the cereal crops are ripe for the sickle. In most representations of Virgo, therefore, she appears sometimes as Ceres, and sometimes as an angel, with ears of corn in her hand, defined in the heavens by the position of the bright star Spica. On this subject, Admiral Smyth has remarked that "we are told that among the Orientals she was represented as a sunburnt damsel, with an ear of corn in her hand, like a gleaner of the fields; but the Greeks, Romans, and moderns have concurred in depicting her as a winged angel, holding wheat ears, typical of the harvest, which came on in the time of the Greeks as the Sun approached Spica. She forms a conspicuous and extensive asterism, replete with astronomical interest." Virgo is bounded on the east by Libra, on the west by Leo, on the north by Boötes and Coma Berenices, and on the south by Corvus, Crater, and Hydra. The number of stars in Virgo, contained in Ptolemy's catalogue, is thirty-two, but Flamsteed has recorded one hundred and ten, and in Bode's Celestial Atlas four hundred and eleven are inserted.

The most brilliant star in Virgo is Spica, which forms, as we have previously mentioned, almost an equilateral triangle with Denebola in the tail of Leo and Arcturus in Boötes. Spica may also be found by drawing a straight line from Beta Boötis through Arcturus, but it may also be readily distinguished by its isolated appearance. Among other star references a long line drawn from Dubhe, in Ursa Major, through Gamma in the same constellation, will reach Spica. Or a line from Polaris, through Mizar, the star in the middle of the tail of the Great Bear, will also pass through Spica. The following lines will popularly guide the observer to the positions of several bright stars above the horizon at midnight during the spring months :—

> " From the Pole star through Mizar glide
>   With long and rapid flight,
> Descend, and see the Virgin's spike
>   Diffuse its vernal light.
> And mark what glorious forms are made
>   By the gold harvest's ears,
> With Deneb west, Arcturus north,
>   A triangle appears ;
> While to the east a larger still,
>   Th' observant eye will start,
> From Virgo's spike to Gemma bright,
>   And thence to Scorpio's heart."

The principal remaining stars in Virgo can be easily identified between Spica and Denebola. They are, however, generally known only by a Greek letter attached to the name of the constellation. The first of these stars from Spica is Gamma Virginis, the two nearer the zenith are Delta and Epsilon, and the two west of Gamma are Eta and Beta. The last star is below Denebola.

By far the most interesting object to astronomers in Virgo is acknowledged to be that extraordinary binary system known as Gamma Virginis. Its position can be identified in the heavens from the preceding explanation. This class of stars, above all others, exhibits to us the proof of the law of gravitation being as applicable in remote regions of the universe as in the comparatively smaller interval of space occupied by the members of our own solar system. It will not be out of place to remark here that double stars are very common in all directions of the heavens, that there is scarcely a constellation in which several are not to be found, and that the number of these objects catalogued by different observers amounts to several thousands. But ordinary double stars must not be confounded with those which have been proved by observation to belong to a common system, for many of the former are known to be only optically double. For example, two stars appearing to the naked eye as one object, but through a telescope as two, may be separated from each other by a distance as great as between any two stars in the heavens, though by accident they are viewed from the Earth in the same line of direction. These apparently double stars are consequently always observed in the same relative order, so that their telescopic measures of distance and angular position remain for ages without sensible alteration. But in the double stars known to be physically connected—hence their name of binary stars— these measures of distance and angular position are always changing more or less, and when observations are made at different epochs, the movements of the stars with respect to each other are very evident indeed. Sir John Herschel remarks that " we have the same evidence of their rotations about each other that we have of those of Uranus and Neptune about the Sun; and the correspondence between their calculated and observed places in such very elongated ellipses, must be admitted to carry with it proof of the prevalence of the Newtonian law of gravity in their systems, of the very same nature and cogency as that of the calculated and observed places of comets round the central body of our own."

As an example of a binary star, we could not select a more appropriate one for our purpose than Gamma Virginis, because it is one which has received constant attention since the beginning of the eighteenth century. When Mayer observed it in 1756, the distance between the two stars was found to be six and a half seconds of arc. Sir William Herschel, in 1780, observed this space to be one second smaller. From recorded observations since that time, the stars have been seen to approach each other gradually, till at length, in 1836, they were so close that the highest magnifying power, applied to the most celebrated telescopes, was unable to separate the two components. After this the star gradually opened, and in 1837 was again seen double when viewed through a good telescope. In 1840 the distance between the components was observed by the Rev. W. R. Dawes, who found it nearly a second and a half; in 1852, from observations made at the Royal Observatory, this distance had increased to upwards of three seconds; and from some excellent measures made by the Rev. R. Main, at Oxford, in 1864, the space between the two stars was equal to four seconds and a quarter. At the present time it slightly exceeds this quantity. In all probability in a few years hence the relative appearance of the stars in Gamma Virginis will be similar to that first recorded by Bradley in 1718, since which time one complete revolution will then have been made. This period, computed from the observations, is about 180 years. In the small diagram of Gamma Virginis we have given

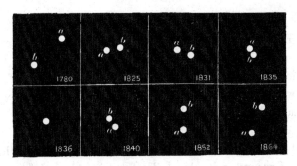

TELESCOPIC VIEW OF GAMMA VIRGINIS IN DIFFERENT YEARS.

a selected number of the telescopic appearances of this beautiful star, which will possibly give a better idea of the relative movements of the components than by any further detailed description.

LIBRA.

Libra, the Balance, is the first autumnal sign, and the seventh in order of the twelve signs of the zodiac. This constellation is bounded on the east by Scorpio, on the south by Centaurus and Lupus, on the west by Virgo, and on the north principally by Serpens. According to Ptolemy, it contained only seventeen stars visible to the naked eye; but in the atlas of Bode, 180 are inserted. The position of the two scales is pointed out by the two principal stars, Alpha and Beta, the former being exactly midway between Spica and Antares. Alpha Libræ is of the third magnitude, and of a pale yellow colour, preceded by a star of the sixth magnitude. Beta Libræ, a pale emerald-coloured star, is of the second magnitude, situated a short distance to the north-east of Alpha. The Balance was considered of old to typify the equality of the autumnal days and nights, as well as the general uniformity of temperature at that season of the year. The sign of Libra has been the subject of a difference of opinion among astrologers, some of whom have placed it among their lucky signs, while others have classed it, owing to its proximity to Scorpio, among those least beneficial to human interests. An illustration of the latter has been gathered from an old illuminated almanack bearing the date of 1386, in which it is calmly stated that " whoso es born in yat syne sal be an ille doar and a traytor." Libra contains several interesting double and triple stars, and two clusters. One of the clusters, No. 5, Messier Libræ, is a beautiful object over the beam of the Balance. Through telescopes fitted with a low magnifying power, this superb cluster has the appearance of a round nebula. When Messier observed it first in 1764, he described it as such, adding the remark, "I am certain that it contains no star." But when Sir William Herschel, in May, 1791, directed his great forty-foot reflecting telescope to it, he found it resolved into separate stars, of which he counted no fewer than 200. At the same time, the central mass was so compressed, that he was not able to resolve that part of the cluster, so as to distinguish the different components.

SCORPIO.

Scorpio, the Scorpion, is the eighth sign in order of the zodiac. Of its reputed origin we have been informed by the ancient poets of Greece, that the Scorpion was sent by Diana to destroy Orion for interfering with the duties of her office. Ovid, however, tells us "that this Scorpion was produced by the Earth, to punish Orion's vanity for having boasted that there was not on the terraqueous globe any animal which he could not conquer." The autumnal season of the year has also been fitly represented by the Scorpion; for whereas the former produces in abundance all kinds of fruits which are frequently the parents of many diseases, so the latter, as he recedes on his path, is supposed to inflict all manner of wounds with his tail. Scorpio is bounded on the east by Sagittarius, on the south by Lupus, Norma, and Ara, on the west by Libra, and on the north by Ophiuchus and Serpens. Antares, called also Cor Scorpii (the heart of the scorpion), is the chief star in this constellation, and is a small first-magnitude star, preceded by a very close companion of a bluish colour. Antares shines with a deep red light, and may be found readily by drawing a line from Vega, through Ras Alague.

"Through Ras Alague, Vega's beams direct th' inquiring eye,
Where Scorpio's heart, Antares, decks the southern summer sky."

Antares, with Aldebaran in Taurus, Regulus in Leo, and Fomalhaut in Piscis Australis, were looked upon by the ancient Persians as the guardian stars of the heavens, dividing the celestial sphere into four equal parts. When Aldebaran was in the vernal equinox, and the guardian of the eastern sky, Antares was in the autumnal equinox, with a like charge of the western sky. Regulus being near the summer, and Fomalhaut the winter solstice, these two stars overlooked the northern and southern portions of the heavens respectively. At the present date, these four stars no longer hold these prominent offices, as the equinoxes and solstices are now in very different parts of the heavens, on account of their retrogression, produced by what is technically called "the precession of the equinoxes."

Scorpio is not a large, but it is a very brilliant constellation, especially when viewed from places south of the equator. Besides Antares, it contains Beta Scorpii, or Iklil, a star of the second magnitude, with an interesting close companion of the sixth magnitude. Scorpio also includes a great number of stars from the third to the fifth magnitudes. This general brilliancy was certain to attract the attention of the astrologers and soothsayers of the early and middle ages. On this subject, the late Admiral Smyth remarks, in his *Celestial Cycle*, that Scorpio was always looked at as a group of stars portending universal evil. He says "Scorpio attracted much notice from the corps of astrologers, with whom it was the 'accursed constellation,' and the baleful source of war and discord; for, besides its being accompanied by tempests when setting, it was of the watery triplicity, and the stinging symbol of autumnal diseases, as it winds along with its receding tail. But though stigmatised as 'the false sign' by seers of every degree, the redoubtable Gadbury, at whose birth it ascended, broke many a lance in its defence, and stoutly contended for its beneficial influences; and the alchymists were well assured that the transmutation of iron into gold could only be performed when the Sun was in that sign."

SAGITTARIUS.

Sagittarius, the Archer, is the third of the southern signs, and the ninth in the order of the zodiac. It can be recognised by eight stars forming two similar quadrangles, one of which is in the Milky Way. From their low altitude, however, they can only be seen distinctly on very clear nights, when near the meridian. In the middle ages, when the influence of astrologers was so great, and when no important undertaking was commenced without a previous consultation with the aspect of the stars, Sagittarius was generally considered a lucky sign. One of these old astrological doctors, by name Arcandum, who published a book in 1542, which was "ryght pleasaunte to reade," has declared that a person born under the sign Sagittarius, "is to be thrice wedded, to be very fond of vegetables, to become a matchless tailor, and to have three special illnesses;" but as the last attack of sickness is to befall the patient at eighty years of age, it is not of paramount moment. Some of our readers will probably question the value of these peculiar advantages so carefully chronicled by the old astrologer. Ptolemy's catalogue contains thirty-one stars in Sagittarius, that of Flamsteed sixty-nine, and the Atlas of Bode three hundred and thirty-nine. The locality of Sagittarius in the heavens can be discovered by drawing a line from Deneb through Altair, which when produced will pass through the centre of the constellation.

"From Deneb in the stately Swan, describe a line south-west,
Through bright Altair in Aquila 'twill strike the Archer's breast."

In the southern hemisphere, in latitude 34°, this constellation attains a very elevated position, most of its stars being, when at their highest point, only a few degrees north of the zenith.

CAPRICORNUS.

Capricornus, the Goat, joins Sagittarius, and is the fourth of the southern signs, and tenth in the order of the zodiac. It is bounded by Sagittarius on the west, Aquila on the north, Aquarius on the north-west and west, and Piscis Australis and Microscopium on the south. Alpha and Beta Capricorni are the two chief stars in this constellation, but they are rather below the third magnitude. They are situated near together, south of the three principal stars in Aquila. Like Sagittarius, the old astrologers always looked with favourable eyes on Capricornus as a lucky sign; for "whoso es borne in Capcorn schal be ryche and wel lufyd." On this subject, the late Admiral Smyth has remarked that "although Capricornus is not a striking object, it has been the very pet of all constellations with astrologers, having been the fortunate sign under which Augustus and Vespasian were born, who thereby were entitled to the tutelage of Vesta; and this Sabæan superstition was honoured by medals, marbles, poems, and what not. It was not only of happy influence in classic times, but was also mightily looked to by the Arabians, who termed Alpha and Beta the lucky stars of the slaughterer, and Gamma and Delta the fortunate stars bringing good tidings." The position of Capricornus can be determined by drawing a line from Vega to the horizon, through Altair, when it will pass between Alpha and Beta Capricorni in the head of the Goat.

AQUARIUS.

Aquarius, the Water-bearer, is not a conspicuous constellation, having no star greater than the third magnitude. It contains, however, its full average of double stars, clusters, and nebulæ. Aquarius is the eleventh of the zodiacal signs, and is situated south of Equuleus and Pegasus, east of Capricornus and Aquila, north of Piscis Australis and Sculptor, and west of Pisces and Cetus. Aquarius was another favourite sign with the old astrologers, who declared that its stars possessed so much influence, virtue, and efficacy, that the seasons were affected by them "in a wonderful, strange, and secret manner;" and, according to an old manuscript almanack for the year 1386, "it es gode to byg castellis, and to wed, and to lat blode" when the Sun is in this sign. The principal object in Aquarius is Alpha Aquarii, or Sadalmelik, the king's lucky star. A line drawn from Alpherat to Markab, and then continued towards the south-west, will pass near Alpha Aquarii.

> "From Scorpio, to where Aries shines, you catch no brilliant ray,
> Through twice two interjacent signs, to mark your trackless way;
> Yet would you know where, from his urn, Aquarius pours the stream,
> From fair Andromeda descend, o'er Markab's friendly beam,
> Or from bright Vega cast your glance, and through the Dolphin's space,
> Then just as far again you'll find the Water-bearer's place."

PISCES.

Pisces, the Fishes, is the twelfth and last of the zodiacal signs, and that in which the Sun crosses the equator at the vernal equinox. It is represented on celestial globes and maps as two fishes widely separated from each other, with their tails joined together by a long ribbon or string. This constellation occupies a considerable space in the heavens, one of the fishes being situated under the right arm of Andromeda, and the other under the wing of Pegasus. A good guide to the position of the two fishes may be obtained by reference to the four stars in the square of Pegasus, a line drawn from Alpherat to Gamma Pegasi being parallel to the body of one fish, while another line from Gamma Pegasi to Markab is likewise parallel to the other fish. Pisces is bounded by Andromeda on the north, by Aries and Triangulum on the east, by Cetus on the south, and by Aquarius and Pegasus on the west. Alpha Piscium, of the third and a half magnitude, is a close double star, and the largest in this constellation. The colours of the two components of this beautiful object are a pale green and blue. About forty stars are visible to the naked eye in Pisces, of which Ptolemy recorded the approximate positions of thirty-eight. Bode's Atlas contains two hundred and fifty-seven. In addition to Alpha Piscium, there are several double stars included within the boundaries of Pisces; but although some of them are of an interesting character, there is nothing unusually remarkable in their appearance or history to require any special notice.

143

CONSTELLATIONS NORTH OF THE ZODIAC.

———◆———

" Take the glass,
And search the skies. The opening skies pour down
Upon your gaze thick showers of sparkling fire;
Stars, crowded, thronged, in regions so remote,
That their swift beams—the swiftest things that be—
Have travelled centuries on their flight to Earth.
Earth, Sun, and nearer constellations ! what
Are ye amid this infinite extent
And multitude of God's most infinite works ?"

HENRY WARE.

———◆———

ANDROMEDA.

ANDROMEDA is situated in a favourable position for observation in the latitude of Great Britain, where that portion of the constellation nearest the pole never sets, the remainder being also above the horizon during a considerable part of the year. In celestial maps, the figure of Andromeda is placed near those of her father, mother, and lover, Cepheus, Cassiopeia, and Perseus. She is generally represented in the bonds which, according to the heathen mythology, she carried with her to the stars. Pegasus and Lacerta are on the west of Andromeda, Pisces on the south, Cepheus on the north-west, Cassiopeia on the north, and Perseus and Triangulum on the east. There are two hundred and twenty-six stars in Andromeda inserted in Bode's Atlas, about thirty of which are contained in the most ancient catalogues. The principal star of this constellation is Alpherat, or Alpha Andromedæ, the north-eastern star of the square of Pegasus. This object heads the list of standard stars in the Nautical Almanac, whose accurate right ascensions and declinations are given for every ten days throughout the year, to be used for the determination of the error of the clock, and for other delicate astronomical purposes. The second star in magnitude is Beta Andromedæ, or Mirach, formerly placed by the Arabian astronomers in the northern fish's head, from which it is now slightly removed. Mirach is of the second magnitude, and of a fine yellow colour. Delta Andromedæ is about midway between the two chief stars Alpherat and Mirach. The situation of the latter star can be pointed out by drawing a line from Alpha Ceti, or Menkar, through the two bright stars in Aries.

144

Or if we follow the directions of the poet, its position in the sky will be very evident, bearing in mind that Alpherat is in the lady's head, and that Markab is the south-west star in the square of Pegasus.

> "From Markab run a line beneath th' imprison'd lady's head,
> And over *Delta* on her back to Mirach 'twill be led."

Gamma Andromedæ is a beautiful triple star in the lady's ankle, and is a very favourite object for amateur observation. This star was first seen double by C. Mayer, in 1778. The principal component is of the third magnitude, and of an orange-yellow colour, while the smaller one is between the fifth and sixth magnitudes, and of a bluish colour. In 1842, M. Struve, of the Pulkowa Observatory, announced that, with his large equatorial, he found the smaller component actually to consist of two stars. This discovery has since been verified by other astronomers, so that this object is now looked upon as one of the severest tests of the space-penetrating power of an astronomical telescope. It requires, however, the use of a superior instrument, with an object-glass of large aperture, to separate the smaller star into two.

Andromeda contains about seventy stars visible without the aid of a telescope, including all down to the seventh magnitude.

AQUILA AND ANTINOUS.

In the midnight sky of summer, Aquila, the Eagle, occupies a very prominent position. This constellation is really composed of two, Aquila proper and Antinous, but it has been found convenient of late years to consider all the stars included in both as belonging to one asterism. In modern catalogues they are known, therefore, only under the name of Aquila. Antinous is situated directly south of Aquila proper, and includes most of the stars between Beta Aquilæ and Sagittarius. The combined constellation is bounded chiefly by Sagitta on the north, Delphinus and Aquarius on the east, Sagittarius on the south, and Ophiuchus on the west. Its position on summer nights, or autumn evenings, can be easily found by its three principal stars, Alpha, Beta, and Gamma Aquilæ, respectively named by the ancients Altair, Alshain, and Tarazed. These three stars have frequently been taken for the three gems in the belt of Orion, to which, however, they bear but a slight resemblance. Altair, a small first-magnitude and the central one, is of a pale yellow colour; Gamma, or Tarazed, the upper star, is of the third magnitude; and Beta, or Alshain, is about a half-magnitude smaller than Gamma. A line drawn through these stars will point in a northerly direction nearly to Vega, and southerly to Alpha and Beta Capricorni. Ptolemy includes in his *Almagest* fifteen stars in Aquila. In the seventeenth century, Hevelius recorded the positions of forty-two; in the early part of the eighteenth century, Flamsteed increased the number to seventy-one; while Bode in his Atlas, published at a much later date, has inserted two hundred and seventy-six.

The Milky Way passes through Aquila, where it crosses the heavens in two distinct branches, the three principal stars being near the eastern edge of the eastern branch.

AURIGA.

Auriga, the Charioteer, one of the ancient asterisms, is generally represented on celestial maps holding a goat and two kids in his left hand. By the Arabs he was termed the Guardian of the Pleiades. According to the ancient mythology, Auriga was placed, after his death, among the stars, on account of his invention of chariots, and for his skill in the management of horses. The goat and the kids were supposed to have been given a place in the heavens in honour of Amalthœa, a daughter of Melissus, king of Crete, who, with her sister Melissa, fed Jupiter with goats' milk during his infancy. It has also been suggested that Auriga was a scientific representation of the fable handed down to us of Phaëton. Two small stars in the kids, Zeta and Eta Aurigæ, named the Hædi, were regarded in days of yore as having an unfavourable influence on the weather. Callimachus says in an epigram of the Anthologia—

> "Tempt not the winds, forewarn'd of dangers nigh,
> When the kids glitter in the western sky."

Capella, the principal star in this constellation, is situated on the body of the goat, or rather on the left or western shoulder of the Charioteer. At midnight in December it is only a few degrees from the zenith, where it shines with great brilliancy. In summer, at midnight, it is a conspicuous object near the north meridian, at an altitude of about seven or eight degrees, and is consequently always above the horizon of Great Britain. Capella is slightly east of the Milky Way, and occupies the summit of a triangle, the base of which is formed by uniting Alpha Cassiopeiæ and Polaris. The nearest three small stars inserted in the diagram for December south-west of Capella mark the position of the kids. Beta Aurigæ, the second star in this constellation, is on Auriga's right shoulder, and when on the meridian, is about six degrees from the zenith. Auriga is bounded on the north by Camelopardalis, on the east by Lynx and Gemini, on the south by Taurus, and on the west by Perseus. In the catalogues of the ancients, the positions of about fourteen stars were registered. Hevelius, in the seventeenth century, included forty in his *Uranographia*, Flamsteed sixty-six in his *Historia Celestis*, while Bode, by collecting together a smaller class of stars, has inserted two hundred and thirty-nine in his atlas. The stars in this constellation are very easily found by alignment, especially as Capella is such a brilliant isolated object. A long line drawn southward, perpendicularly to the Ursa Major Pointers, will lead to Auriga. But if we look in an upward direction from Orion, we may profitably take the rhymer's advice, Nath being Beta Tauri on the tip of one of the horns of Taurus.

> "From Rigel rise, and lead a line through Bellatrix's light,
> Pass Nath, upon the Bull's north horn, and gain Capella's height—
> Where a large triangle is form'd, isosceles it seems,
> When Beta is with Delta join'd to lustrous Alpha's beams."

BOÖTES.

Boötes is one of the principal constellations of the summer evening sky. It is bounded on the east by Corona Borealis and Serpens, on the south by Virgo, on the west by Canes Venatici and Coma Berenices, and on the north by Draco. In the ancient catalogue of Ptolemy, twenty-three stars were included as belonging to this constellation; in the *Historia Celestis* of Flamsteed, observations of fifty-four have been recorded; while, in the more modern celestial atlas of Bode, three hundred and nineteen are inserted. Arcturus, one of the most brilliant stars in the northern hemisphere, is situated between the legs of Boötes. In ancient times it was a noted star among mariners, who, however, looked upon its influences with suspicion. The squally weather which generally preceded the autumnal season was ascribed by the Greeks to the power of Arcturus. It is related by Demosthenes "that a sum of money was lent at Athens on a vessel going to the Crimea and back, at the rate of $22\frac{1}{2}$ per cent., with the understanding that unless the ship returned before the rising of Arcturus, 30 per cent. was to be paid."

Arcturus was the first star observed with a telescope in daylight. This feat was announced in 1635 by a M. Morin. It appears, however, that little notice was taken of the occurrence till the year 1669, when the Abbé Picard published the results of an observation of Arcturus, made when the Sun was seventeen degrees above the horizon. This discovery created a sensation among astronomers, who were gleaning at this time quite a harvest among the stars by the use of the telescope, lately invented. Owing, however, to the small object-glasses of the first telescopes, which were only toys compared with the magnificent reflectors and refractors of modern days, the astronomer of the seventeenth century was unable to see more than the principal celestial objects. Several important discoveries were, however, made in that century, among which may be mentioned the four satellites of Jupiter and five of those of Saturn, together with the ring of the last-named planet. We have still preserved, on the walls of the Royal Observatory, the transit instrument used by Dr. Halley at that place, with an object-glass no larger than that of many modern ship spy-glasses. If we compare Dr. Halley's small object-glass with one lately constructed for Mr. Newall by Messrs. Cooke, with a clear aperture of twenty-five inches, or with the six-foot speculum in the reflecting telescope of the Earl of Rosse, we shall not be surprised that the astronomers of the seventeenth century appeared somewhat elated at having seen a star in broad daylight. This is now no uncommon occurrence. In the winter days, the author has frequently observed at noon-day, not only such bright stars as Arcturus, Vega, and others, but occasionally some as small as the fourth and fifth magnitude; and on one occasion, when the atmosphere was more than usually pure, a star of the sixth magnitude was observed within two hours of noon. In short, from the beginning of November to the end of February, many of the fourth magnitude are observed on the meridian of Greenwich with the transit-circle at the Royal Observatory.

Boötes is figured generally as a robust man holding in one hand a club, spear, pastoral

147

staff, or sickle; for at various epochs he has been represented with each of these symbols. The other hand is upraised towards Canes Venatici. When Hevelius introduced the Greyhounds in his *Uranographia*, published in 1690, they were attached to Boötes and placed near the hind legs of the Great Bear. Boötes has therefore been called the Bear-keeper, and also the driver of the waggon composed of the seven chief stars in Ursa Major. Arcturus was for some time supposed to be the nearest fixed star to the earth; but there have been many others now found to be much nearer. There is, however, no doubt that it is less distant from us than the great bulk of the stars. Epsilon Boötis, sometimes called Izar, is a double star, interesting for the distinct and contrasting colours of its components; the principal one being of the third magnitude and of a pale orange colour, while the companion is of the seventh magnitude, and of a sea-green colour. This star has been frequently seen double in the daylight, within two hours of noon, although the companion is so small. It is not known at present whether the two stars form a binary system; if so, its period of revolution must extend over not less than a thousand years. Xi Boötis is another interesting object of this class, the principal star being of the third and a half magnitude, and of an orange colour, and the secondary about the sixth or seventh magnitude, and of a purple tint. The contrast between the colours is very brilliant.

The following diagram exhibits a selected list of double stars, as seen in an inverting astronomical telescope. Some of them are, probably, only optically double, their apparent juxtaposition being the result of the two stars being seen from the surface of the earth in the same line of direction, or, if they be binary systems, their time of revolution must extend over a lengthened period. Castor, Epsilon Hydræ, Gamma Leonis, and 61 Cygni, have been proved to belong to the latter class of objects. Gamma Andromedæ is triple, the companion being resolvable into two components when viewed through a good defining telescope, especially when the star is at a high altitude about sunrise or sunset. The list of known binaries exceeds a hundred, without including several double stars suspected to belong to a common gravitational system.

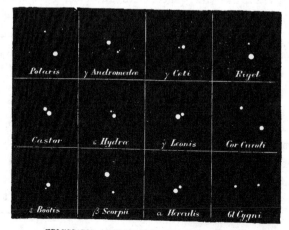

TELESCOPIC APPEARANCE OF DOUBLE STARS.

CAMELOPARDALIS.

Camelopardalis is a modern constellation introduced into the heavens by Hevelius in the seventeenth century. It occupies a very considerable space in the sky north of the zenith, between Auriga and Ursa Minor, and contains a large number of small stars, only four of which are equal to the fourth magnitude. Camelopardalis has Ursa Major and Lynx on the east, Perseus and Auriga on the south, Perseus and Cassiopeia on the west, and Cepheus, Draco, and Ursa Minor on the north. This constellation, combined with Lynx, gives a very bare appearance to a large portion of the circumpolar sky between Polaris and Auriga and Gemini.

CANES VENATICI.

Canes Venatici, or the greyhounds Chara and Asterion, form a small constellation of comparatively recent origin, having been introduced into the heavens by Hevelius, in the seventeenth century. It is situated north of Coma Berenices, and occupies an empty space between Boötes and the hind legs of Ursa Major. Excepting its principal star Cor Caroli (Charles's heart), there is no object worthy of special notice in this constellation. In most celestial atlases or globes, Cor Caroli is placed in the centre of a heart attached to a crown on the shoulders of Chara. It is a beautiful double star, the larger component being white, and the smaller a pale lilac colour. The following anecdote of the origin of the name of the star is given by Admiral Smyth: "But it came to pass that it was named Cor Caroli by Halley, at the suggestion of Sir C. Scarborough. The popular story, or rather the vulgar one, runs—how Scarborough, the court physician, gazed upon a star the very evening before the return of King Charles II. to London, the which, as in duty bound, appeared more visible and refulgent than heretofore; so the said star, which Hevelius had already made the lucida of Chara's collar, was thereupon extra-constellated within a sort of valentine figure of a heart, with a royal crown upon it; and so the monarch, it would seem by this extraction, remained heartless." Cor Caroli can be readily identified: a line drawn from Polaris through the first star in the tail of Ursa Major will lead directly to it. If we take advantage of the rhymester, we shall obtain other directions for finding not only Cor Caroli, but also the important group of Corona Borealis.

> " When clear aloft, Boötes seek,
>   His brilliance leads the gaze,
> And on each side its glitt'ring gems
>   The spacious arch displays ;
> Arcturus east to Vega join,
>   The Northern Crown you'll spy ;
> But west, to Ursa's second star,
>   He marks Cor Caroli."

Cor Caroli also forms an equilateral triangle with Gamma Ursæ Majoris and Alkaid in the hip and tail of the Great Bear.

CASSIOPEIA.

The name of Cassiopeia has been derived from the wife of Cepheus, and the mother of Andromeda. This constellation is one of the most attractive groups of stars visible in the sky of the northern hemisphere, and contains several stars of the second and third magnitudes. It is always in direct opposition to Ursa Major, with respect to the Pole. For example, if Cassiopeia be at its greatest elevation, Ursa Major is near the horizon in the north, and *vice versâ*. If the Great Bear be in the east, Cassiopeia is in the west, and so on throughout their diurnal revolution. It is not difficult, therefore, to find Cassiopeia; for a line drawn from the middle of Charles's Wain, through Polaris, passes across the centre of the group. Cassiopeia is bounded by Cepheus, Perseus, Andromeda, and Camelopardalis, and when passing the meridian above the Pole, it is directly overhead in all parts of the British Isles. The principal stars bear some resemblance to the letter Y with the vertical stem a little bent; or, by viewing it from another direction, to a badly-formed W; they have, however, frequently been called Cassiopeia's, or the lady's, chair, imagination having transformed these five or six stars into the form of an antique seat.

POSITION OF THE TEMPORARY STAR OF 1572.

Cassiopeia is celebrated as being the spot where a very remarkable temporary star appeared in November, 1572. It suddenly burst forth with a brilliancy greater than any star around it, and was therefore noticed by several persons about the same time. Tycho Brahé was one of these fortunate observers. From his own account, we gather that he was returning from his laboratory, on the evening of November 11th, when he saw a group of peasants gazing at a brilliant object, which he knew had not existed in that shape an hour previously. It was at first supposed to be a comet, though it had no distinctive marks of being one, but twinkled like any ordinary fixed star. It increased rapidly in magnitude, till it surpassed Sirius and Jupiter in lustre, and was even observed at noonday. This great brilliancy, however, continued only for a short time, when it gradually diminished. In March, 1574, it became invisible to the naked eye, and has not been observed since. A severe scrutiny of the neighbourhood in which this wonderful star

150

appeared has been made since the construction of the powerful astronomical instruments of modern days, but no trace of it can be found. La Place, the celebrated French mathematician and astronomer, was led to believe that the frequent change of colour which was observed in this star, first white, then yellow, afterwards reddish, and finally a bluish tint, showed that the sudden blazing forth was probably caused by the action of fire. This hypothesis was not, however, generally believed at the time. But since the sudden outburst of the star in Corona Borealis, which is noticed in our description of that constellation, many astronomers of the present day are inclined to believe that La Place's suggestion is not so improbable as it first appeared.

CEPHEUS.

According to the ancients, the constellation Cepheus was placed in the heavens in memory of a king of Ethiopia, or India, husband of Cassiopeia, and father of Andromeda. The old Ethiopic name of this asterism was Hyk, a king. Cepheus is bounded on the north by Ursa Minor and Camelopardalis, on the west by Draco, on the east by Cassiopeia, and on the south by Lacerta and Cygnus. The stars Alpha, Beta, and Gamma Cephei, form an arc, of which Beta Cassiopeiæ is nearly the centre. The head of Cepheus is situated in the Milky Way, and can be identified in the August diagram, looking north, near the zenith, by three stars of the fourth magnitude forming a small triangle. The chief objects in Cepheus are Alderamin in the right shoulder, Beta, or Alphirk, in the waist, and Gamma, according to Ptolemy, in the left foot. Cepheus was one of the old forty-eight asterisms, and of considerable note among the wandering shepherds of Arabia. It contains several choice and remarkable double stars, nebulæ, and clusters, which have afforded ample subjects for the scrutinizing eyes of modern astronomers. The position of Cepheus and its principal stars can be clearly recognised between Polaris and the zenith, having the bright group of Cassiopeia on the east, and the chief stars of Draco on the west. We can almost imagine the exact form of this circumpolar constellation from the following lines :—

"Near to his wife and daughter, see aloft where Cepheus shines.
The wife, the Little Bear, and Swan, with Draco bound his lines ;
Above Polaris, twelve degrees, two stars the eye will meet,
Gamma, the nomade shepherd's gem, and Kappa—mark his feet ;
Alphirk, the Hindu's Kalpeny, points out the monarch's waist,
While Alderamin, beaming bright, is on the shoulder placed :
And where o'er regions rich and vast, the Via Lactea's led,
Three stars, of magnitude the fourth, adorn the Ethiop's head."

COMA BERENICES.

This small constellation contains about forty stars visible without the aid of a telescope, several of which are of the fourth and fifth magnitudes. It is situated south of Canes Venatici, west of Boötes, north of Leo and Virgo, and east of Leo and Ursa Major. Its name, Berenice's Hair, has been derived from an Egyptian fable. It has been asserted that Evergetes, a king of Egypt, was absent for some time on a dangerous expedition, so long indeed as to cause considerable anxiety in the mind of the Queen Berenice. She consequently made a vow to consecrate her fine head of hair to the goddess Venus if her husband were permitted to return in safety. On the successful return of the king shortly afterwards, she fulfilled her vow by at once ordering her locks of hair to be cut off and to be hung up in the temple of Venus. Their beauty, however, appears to have attracted the attention of the gods, for the tresses very soon mysteriously disappeared. Recourse was had to the wise men of the period for an explanation, when one Conon, an astronomer, declared that they were taken by Jupiter, who turned them into a constellation of stars. This fable, like many others of the same kind, must only be taken for what it is worth; but it is true enough that several of the ancient philosophers have alluded to these stars as "the tresses." To the naked eye, the principal group has a nebulous, or rather woolly appearance, owing to the aggregation, in a limited space, of a number of stars of the fourth, fifth, and sixth magnitudes. These can be easily found by drawing an imaginary line from the bright star Alkaid on the tip of the tail of the Great Bear, through Cor Caroli, as far as Denebola. About midway between the two last-named stars, the line will pass through the group.

CORONA BOREALIS.

Corona Borealis, or the Northern Crown, is a small but important constellation between Böotes and Hercules. It consists principally of a group of well-known stars in the form of a crescent, the brightest of which is Alphecca, or Gemma, of the second magnitude. This star is placed in the centre of the front of the crown, as the most precious stone in the diadem. Four of the stars are of the fourth magnitude. This constellation, confined as it is within so very limited a space, contains, nevertheless, some interesting objects, including three binary, in addition to other double stars, and also a triple and a quadruple star. Corona Borealis is one of the ancient asterisms, and is supposed to have derived its name as far back as the date of the origin of the zodiacal signs. It was also considered to be the crown of the virgin, from its rising immediately after Virgo. According, however, to the mythology of the Greeks, the origin of the name arose from the traditional presentation by Bacchus of a beautiful crown to Ariadne, daughter of Minos, king of Crete. On the death of Ariadne, who had become the wife of Bacchus, and as a memorial to her honour, this crown was placed among the stars, forming the constellation Corona Borealis. By reference to other stars, there are various ways by which this popular group can be pointed out. One of them is by passing a line from the last two stars in the tail of the Great Bear, through the northern part of Böotes, when it will pass a short distance south of the group. In the spring months the following alignment in rhyme will also serve this purpose, Epsilon Virginis, referred to in the first line, being the most northerly conspicuous star in Virgo above Spica :—

" From *epsilon* in Virgo's side Arcturus seek, and stem,
  And just as far again you'll spy Corona's beauteous gem ;
  There no mistake can well befall e'en him who little knows,
  For bright and circular the Crown conspicuously glows."

It was near Epsilon Coronæ Borealis, the most easterly star of the principal group, that a most extraordinary temporary star suddenly appeared on May 12th, 1866, shining equal to one of the second magnitude. The first trustworthy observation was made at Tuam, Ireland, but it was on the same and subsequent days discovered independently in various parts of Europe, North America, and India. When first noticed, its lustre equalled Alphecca, but the daily diminution of brightness from the day of discovery was very rapid, amounting for some time to an average rate of half a magnitude each day. Within a month it decreased to the ninth magnitude, to a mere point even when viewed through our principal telescopes. The relative magnitudes, determined by comparison with neighbouring known stars, are as follows : —

May 12	..	..	2 magnitude.
,, 15	..	..	3·5 ,,
,, 18	..	..	4·8 ,,
,, 21	..	..	6·7 ,,
,, 24	..	..	7·8 ,,
,, 30	..	..	8·8 ,,
June 30	..	..	9·0 ,,

M. Ernest Quetelet, of Brussels, remarked that the star, when viewed by the naked eye, decidedly twinkled much more than the other stars near, so much so at times that its variations rendered the observations of its relative brightness extremely difficult.

This wonderful object is all the more curious from the circumstance that it is a small catalogued star observed by M. Argelander, of Bonn, in the years 1855 and 1856, and noted by him as of the ninth and a half magnitude. From the sudden outburst of this star, which was not previously suspected of variability, astronomers have been led to consider that it most probably had been subject to some peculiar catastrophe, the effects of which first became apparent to us in May, 1866. If the outburst consisted of inflammable gas, the star would be supposed to return naturally to its normal size as soon as the gas was completely consumed. From the second week in June till the middle of August, it was recorded of the ninth magnitude ; but in the autumn of 1866, the light of the star again increased for a few weeks, and then again resumed its ordinary appearance, as first observed in 1855 by M. Argelander. It has been frequently examined since, but no appreciable change in its lustre has been noticed. As a proof that something peculiar has been going on in this hitherto comparatively insignificant object, it may be stated that the analysis of its light, as viewed through a spectroscope, has not only exhibited the spectrum with dark absorption lines, similar to those in the spectra of other stars, but, in addition, a series of bright lines has been observed, superposed on the usual spectrum,

153

indicating that the light by which the secondary spectrum was formed, was emitted by matter in a state of luminous gas. The position of one of the bright lines was coincident with that found from the analysis of the light produced from the combustion of hydrogen gas. Many explanations, or rather speculations, have been given concerning the origin of this remarkable outburst, but nothing of a decided nature has been published. It may possibly be a variable star analogous to those which are known to have their regular periods of increase and diminution of lustre ; but still its peculiar double spectrum must naturally lead us to infer that the surface, or more probably the atmosphere, of this distant globe has been subjected to a conflagration of some kind, or, as it has been aptly termed by the Rev. C. Pritchard, formerly president of the Royal Astronomical Society, "the atmosphere of a world on fire." The position in the heavens of this curious object, as well as its relative size with respect to the stars in Corona Borealis, can be readily seen by reference to the accompanying small diagram, in which the new star is inserted at its maximum magnitude, as observed on the night of its discovery.

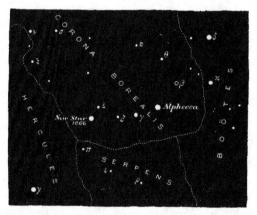

CHART SHOWING THE POSITION OF THE NEW STAR IN CORONA BOREALIS.

A few stars have lately been examined, whose spectra show bright lines very similar in their nature to those of this temporary star in Corona Borealis, and consequently differing considerably from the general stellar spectrum in which black absorption lines only are visible. M. Secchi first pointed out Gamma Cassiopeiæ as one of these abnormal stars, and MM. Wolf and Rayet have since added three others of small magnitude, very near each other in Cygnus, to the list.

CYGNUS.

Cygnus, the Swan, is an important constellation between Lyra and Cepheus, and also one of the ancient asterisms. It is situated in the Milky Way, having Draco and Cepheus on the north, Lacerta and Pegasus on the east, Lyra on the west, and Vulpecula on the south. Its principal star, Deneb, or Alpha Cygni, at the root of the Swan's tail, is of the first magnitude. With Vega and Altair, it forms a well-known stellar triangle, visible in the summer and autumn months. The second star in Cygnus, Albiero, is a double star, celebrated, as we have before observed, for the brilliant contrasting colours of its components, which are respectively of the third and seventh magnitudes. Sir William Herschel made a careful comparison of the relative brightness of the stars in Cygnus, the results of which are to be found in the *Philosophical Transactions* for 1796 and 1797 The ancient catalogue of Ptolemy contained only nineteen stars in Cygnus, but Bode's Atlas has three hundred and sixty. Cygnus has been noted as the locality of two new stars. The first was observed in 1600 by Jansen, Kepler, and others, and continued visible till 1621, when it became too faint to be seen by the naked eye. Cassini, however, saw the star again in 1655, when it was of the third magnitude, but it soon became again invisible. This object is now known to be a periodically variable star, with an interval of eighteen years between its maximum and minimum brightness. The second new star was discovered by Kirch in 1686, in the neck of the Swan. When at its minimum lustre, this object is too faint to be seen with the most powerful telescope.

A small double star, 61 Cygni, the components of which are of nearly equal magnitude, is one of the most interesting objects in this constellation. It is a carefully observed binary system. Its interest, however, principally consists in the large proper motion in the heavens of the two components, from which they have been unitedly regarded as one of the nearest of the fixed stars to our solar system. From some delicate researches of M. Bessel, a German astronomer, this comparatively near star to us has been found to be at a distance of about 650,000 times greater than that between the Earth and the Sun.* Such immense celestial distances are almost beyond the power of the human mind to form a proper conception of their magnitude. Light is known to travel at the rate of about 185,000 miles a second, but, with this great velocity, it would take more than ten years to pass over the space between 61 Cygni and the Earth. When we, therefore, view this star, we are seeing it as it existed ten years ago ; and were any sudden conflagration to take place on its surface to-day, we should only be cognizant of it on our globe in ten years from this date. But this is one of the nearest of the fixed stars ; others are supposed to be a thousand times more distant, or even so far that their light has not reached us since their creation. The large and equal proper motion in space of the two stars composing 61 Cygni is a convincing proof of their physical connection, independently of any change in their relative positions.

Deneb is such a prominent object south of the zenith in the summer night-sky, that it seems scarcely necessary to give any special rule for finding this constellation. But if we look at the square of Pegasus, or rather at Alpherat, in the north-east corner of the square, a line from that star to Vega will nearly pass through Deneb.

---

* The distance between the Sun and the Earth, from the most recent determinations, is, in round numbers, ninety-two millions of miles.

" From the wing's tip, Alpherat through,
   Now skim aslant the skies,
And lo ! bedeck'd with glorious stars,
   The soaring Cygnus flies :
Or, from the westward should you wish
   The same to gaze upon,
Arcturus, Gemma, Vega, join
   To lead you to the Swan."

DELPHINUS.

Delphinus, the Dolphin, although a small constellation, is one of the ancient forty-eight asterisms. The chief stars can be seen in the diagrams as a group east of the three stars in Aquila. It is bounded on the north by Vulpecula, on the west by Aquila, on the east by Pegasus, and on the south by Aquarius. According to Ptolemy, Delphinus contained ten stars, but Hevelius in the seventeenth century increased the number to fourteen. Bode, with telescopic aid, registered fifty-one. Seven of the stars are above the sixth magnitude. Kaswini relates that the lowest star of the group is called the dolphin's tail, and the four in the middle, the necklace or sometimes the cross, that in the tail being the stem of the cross. These stars can be found easily as above, or by the following alignment. Draw a line from Beta and Gamma Lyræ, below Vega, through Albiero, or Beta Cygni, when, if continued twice that distance, it will pass through the centre of the group. Or, according to the *Celestial Cycle*, it can be pointed out by the following rhymes :—

" To heaven's grand arch from deepest seas,
   Behold the Dolphin rise,
The grace, as old Manilius saith,
   Of ocean and the skies :
'Tis placed between that space wherein
   The eagle's wings are spread,
And those few stars unto the east
   Which mark the horse's head."

DRACO.

Draco, the Dragon, is usually represented as surrounding the north pole of the ecliptic, its tail dividing the two Bears, while its head reaches to the right foot of Hercules. It was one of the original forty-eight constellations, and contains, according to Bode's Atlas, two hundred and fifty-five stars. Alpha Draconis, or Thuban, was formerly recorded as the brightest star in Draco : it is now, however, only between the third and fourth magnitude. Upwards of 4600 years ago, this star was situated very near to the pole of the heavens. In the times of the Chaldeans, when the birth of astronomical science is supposed to have taken place, Alpha Draconis, being at that time so near to the celestial pole, must have appeared even more stationary than the pole-star of the present age. It has now deviated from that position nearly twenty-five degrees. Beta and Gamma, of the second magnitude, are the two brightest stars in Draco. The zenith-distance of Gamma Draconis is daily observed at Greenwich, if the weather be favourable, owing to its passing across the upper meridian nearly in the zenith of the Royal Observatory. From observations of this kind made at Wanstead with a peculiar instrument called a zenith sector, Dr. Bradley, formerly Astronomer Royal, made the important discovery of the aberration of light. The observations of this standard Greenwich star are continued from year to year with the object of obtaining fresh data for the re-determination, with the greatest attainable accuracy, of the value of this and other astronomical constants. The advantages arising from these zenithal observations are very great; for when objects culminate in that point, the effects of refraction produced by the Earth's atmosphere are reduced to a minimum. We have previously remarked that the pole of the ecliptic is situated in Draco, and that the nearest star to it greater than the fourth magnitude is Zeta Draconis. Some conception of the general extent of this constellation can be gathered from the following lines :—

"A line from Dubhe, in the Bear, sent right the Guards between,
The stars which form the Dragon's tail in midway will be seen.
Far to the east the body winds, where Lyra's lustres glow,
A ray from Vega to the Pole its lozenge-head will show."

Although Draco contains no first-class star, yet few northern constellations are richer in second-class objects. Many of them can be identified on any clear night between Charles's Wain and the Lesser Bear.

EQUULEUS.

Equuleus, the Little Horse, although of small extent, is one of the original forty-eight constellations of Ptolemy and his predecessors. It does not appear in the astronomical writings of Aratus and Eratosthenes, and its origin is unknown. Thomas Hood, writing in the sixteenth century, remarks that "this constellation was named of almost no writer, saving Ptolomee and Alfonsus, who followeth Ptolomee, and therefore no certain tail or historie is delivered thereof, by what means it came into heaven." Its position may be distinguished by a trapezium of four stars of the fourth magnitude preceding Epsilon Pegasi. These were all the stars in Equuleus known to Ptolemy, Tycho Brahé, and Kepler. Flamsteed recorded the positions of ten, and Bode has inserted thirty-six in his atlas. The surrounding asterisms are Delphinus on the north and west, Aquarius on the south, and Pegasus on the east and north. The head being the only part of the animal represented in celestial maps, it has sometimes received the name of *Equi Sectio*.

HERCULES.

Hercules is a constellation of great extent and importance, although it contains no star greater than the third magnitude. It joins Draco on the north, Lyra on the east, Ophiuchus on the south, and Serpens and Corona Borealis on the west. By the absence of stars of the first and second magnitudes, Hercules is not so remarkable to the naked eye as many other asterisms, but, telescopically, it is a very interesting constellation on account of

157

the numerous double stars, clusters, and nebulæ contained within its boundaries. In the star-catalogues of Ptolemy, Tycho Brahé, and other astronomers who flourished before the invention of the telescope, about thirty stars in Hercules have been inserted, all of which are observable without optical aid. In Bode's Atlas, four hundred and fifty-one are included, a considerable number of which are telescopic. The principal star is Alpha Herculis, or Rasalgeti, one of the most lovely double stars in the heavens; the chief component being about the third magnitude and of an orange colour, while the companion, which is very close, is of the sixth magnitude and of an emerald or bluish-green colour. Rasalgeti varies in brightness to a small extent. In the diagrams it can be noticed a short distance north-west of Ras Alague, or Alpha Ophiuchi. As Ras Alague, Altair, and Vega nearly form an equilateral triangle, the position of Rasalgeti as well as that of Ras Alague can be easily found, each star being respectively in the heads of Hercules and Ophiuchus.

> "Amid yon glorious starry host, that feeds both sight and mind,
> Would you the Serpent-bearer's head and that of Herc'les find,
> From Altair west direct a ray to where Arcturus glows,
> One-third that distance, by the eye, will both these heads disclose."

Among the numerous double stars in Hercules, the most interesting is Zeta, of the third magnitude. The companion is of the sixth. Zeta Herculis is situated near Epsilon, of the same size, between Alphecca in Corona Borealis, and Vega. It is a very close binary, and for many years appears only as one star, even in superior telescopes. Its duplicity was first discovered by Sir William Herschel, in July, 1782. Thirteen years after, he again saw it as a double star; but within a short period of that date it became apparently a single object of a wedge-like form, when viewed on favourable occasions with the great forty-foot reflecting telescope. The disappearance of the companion was considered by Sir William Herschel to be so curious that he made a remark at the time that "the observations of this star furnish us with a phenomenon which is new in astronomy; it is the occultation of one star by another." Zeta Herculis was again seen double in 1826, in which year it was observed at Dorpat by Struve. In 1840, it was easily separable in ordinary telescopes. In 1863 Mr. Dawes observed it as a single object; but in 1865, with his eight-inch refractor, he again saw the small star perfectly detached from the large one.

LACERTA.

Lacerta, the Lizard, is one of the small constellations formed by Hevelius at Dantzig, in the latter part of the seventeenth century. It is bounded on the north by Cepheus, on the east by Andromeda, on the south by Pegasus, and on the west by Cygnus. It contains no star greater than the fourth magnitude. In the June diagram, looking north, the position of Lacerta is indicated by six small stars at the right edge of the Milky Way, between Cepheus and Beta Pegasi.

LEO MINOR.

Leo Minor, the Lesser Lion, was also formed by Hevelius about the year 1691, out of the district between Leo and Ursa Major. In his revision of the constellations, he selected this place for the habitation of the Lesser Lion, for "since they esteem the Bear and the Lion as the hottest and fiercest animals, I wished to place there some quadruped of the same nature." The group of small stars inserted in the diagrams, looking south, between the Sickle in Leo and Lambda and Mu Ursæ Majoris, belong to Leo Minor.

LYNX.

Lynx is another of the new constellations formed by Hevelius out of the stars unappropriated by the ancients. It occupies an extensive space between Auriga and Ursa Major, but it contains no object greater than the fourth magnitude, while the great majority of them scarcely exceeds the sixth. Lynx has Ursa Major and Camelopardalis on the north, Leo Minor on the east, Cancer and Gemini on the south, and Auriga and Camelopardalis on the west. A considerable portion of this constellation is circumpolar in Great Britain.

LYRA.

Lyra, the Harp, one of the old forty-eight constellations, is of very limited extent, but it contains, nevertheless, several important objects, including Vega, one of the brightest stars in the northern hemisphere. Lyra is situated to the south-east of the head of Draco, having Hercules on the west and south, and Cygnus on the east. This small constellation contains about twenty stars visible to the naked eye, but Bode has included one hundred and sixty-six in his celestial atlas, the majority being of course telescopic. The isolated position of Vega, with respect to other very large stars, is so well known that it is scarcely necessary to give any special alignment. It may be, however, as well to state that a line drawn from Arcturus through Corona Borealis leads directly to the Swan, and in its course passes over Lyra. A large right-angled triangle is also nearly formed by the stars Vega, Arcturus, and Polaris. Also the three stars in the neck of Aquila point directly to it. Vega is accompanied by a small telescopic star at a short distance, and some minute objects have been suspected to be very close to the principal star in whose rays they are lost. According to some experiments made by M. Laugier on the relative intensity of the light of the stars, that of Vega was found to be greatly inferior to that of Sirius, although it was placed by him second on the list. M. Siedel, however, from similar experiments, but by a different method of investigation, placed it third in order of lustre. Dr. Wollaston had previously published the results of his researches on the comparative intensity of the light of the sun, moon, and stars, from which he concluded that the intrinsic brightness of Vega is only one-ninth part of that of Sirius. Sir John Herschel, including the stars in both hemispheres, places the relative brightness of the principal stars in the following order:—Sirius, Canopus, Alpha Centauri, Arcturus, Rigel, and Vega. Of these, Canopus and Alpha Centauri are not visible in this country. When near the

159

meridian, Vega certainly appears intensely brilliant through a telescope, at which time it forms a great contrast to the humble light of its small companion. This star, valuable as it is at present to the navigator, will be infinitely more so in future ages, when it will be, both in position and lustre, the polar gem of the northern hemisphere. It will then be apparently stationary, like our present Polaris, with respect to the horizon. The intensity of the light of Vega is sufficiently great to enable any one to observe it with ordinary telescopes during any part of the day or night throughout the year, excepting only for a very short time when it is in the north horizon, where even it can be occasionally observed on very clear nights. Vega, Deneb, and Altair form the well-known triangle by which the three stars are generally identified in the sky, the three angles being occupied by Vega in the north-west, Deneb in the north-east, and Altair in the south.

Beta Lyræ, a star of the third magnitude, south of Vega, and a little west of Gamma Lyræ, of a similar size, may be found by drawing a line from Vega towards Altair, when it will pass between these two stars. Beta Lyræ is accompanied by a companion of the eighth magnitude, and by two smaller ones at a greater distance. There does not appear to be any evidence of a physical connection between them. The magnitude of Beta has been found to be slightly variable. Epsilon Lyræ is a very curious star, and is what may be termed a double-double star. It is situated on the frame of the Harp, a short distance north-east of Vega when on the meridian. To the unassisted eye it appears of the fifth magnitude, but when viewed through a telescope it is separated into two systems of stars, each system being a fine binary pair. Between the two sets three or four very faint stars can be noticed. Admiral Smyth remarks in his *Celestial Cycle*, that these stars form a fine subject for contemplation; "the two sets resembling each other so closely in magnitude, distance, orbital retrogradation and proper motions, as to afford palpable evidence of their forming a twin system; and a combined rotation about a common centre of gravity may be suspected." When contemplating interesting groups of this kind no observer can really believe that these apparently small objects, which seem to be so dependent upon one another for their uniformity of motion, have been placed in their respective positions in the universe without some high purpose, of which we are profoundly ignorant. One thing we do know is, that they have not the slightest influence on our solar system, nor are they useful to man as ornaments of our skies, especially when the object is so minute as to be only visible through good space-penetrating telescopes. There are several other stars of this class in Lyra, double, triple, and quadruple; but there are none of so interesting a character as the group composing Epsilon Lyræ.

OPHIUCHUS.

Ophiuchus, the Serpent-bearer, is one of the old forty-eight asterisms, and occupies at midnight, in May and June, a very large portion of the sky in the south; but although it is of so great an extent, it contains but few important stars. It is bounded on the north by Hercules, on the east by some small constellations, on the south by Scorpio, and on the west by Serpens. The last-mentioned constellation is, however, so entwined with Ophiuchus that a proper separation of the two is extremely difficult. Ophiuchus includes several clusters and double stars, one of the latter, 70 Ophiuchi, being a most interesting binary system. The components of this star revolve around each other in about 100 years. In 1604, a new star of great brilliancy suddenly appeared in the foot of Ophiuchus, but after shining for some time as bright as one of the first magnitude, it totally disappeared in a few months. At the present time no star within the limit of vision, even with the assistance of our largest astronomical telescopes, can be found near its place. The position in the heavens of Ras Alague, the principal star of this constellation, has been previously pointed out by alignment, but the following couplet may still be useful for its detection among the stars :—

> " From Altair let a ray be cast, where we Arcturus view,
> One-third that distance will reveal the star Ras-al-ague."

Ophiuchus contains about eighty stars visible to the unaided eye ; twenty-nine of these are included in Ptolemy's catalogue in the *Almagest*.

PEGASUS.

Pegasus, the Winged Horse, is a prominent constellation on the meridian at midnight in September. It contains several stars of the third magnitude, in addition to Markab, Scheat, and Algenib, in the square, and, according to Bode, three hundred and ninety-three stars. It is bounded on the north by Lacerta and Andromeda, south by Aquarius, east by Pisces, and west by Equuleus and Delphinus. Alpha Pegasi, or Markab, is a white star of the second magnitude, at the junction of the animal's wing and shoulder, while Beta Pegasi, or Scheat, is a deep yellow star of the second magnitude in the left fore leg. Algenib, or Gamma Pegasi, is situated on the extremity of the horse's wing. Epsilon, in the mouth, and Zeta, in the neck, are two conspicuous stars west of Markab and above Aquarius. Eta Pegasi is near Beta, a little to the north-west.

Pegasus and the adjoining constellations nearer the pole owe their position in the heavens to a mythological connection. The fable runs thus :—To assuage the anger of Neptune, Andromeda, the daughter of Cepheus and Cassiopeia, was bound to a rock to be devoured by a sea monster. At the moment when her death appeared inevitable, Perseus, who was returning through the air from the conquest of the Gorgons, changed the sea monster into a rock by showing him Medusa's head, released Andromeda, and subsequently married her as a reward for his trouble. Pegasus is said to have sprung from the blood of the Gorgon Medusa, after Perseus had cut off her head. These five constellations, representing the names of the principal characters in this fable, occupy in September a considerable portion of the sky both north and south of the zenith. Andromeda and Pegasus are south, Cepheus and Cassiopeia north, and Perseus west of that point. Equuleus, the Little Horse, is situated between Pegasus and Delphinus. The observer ought to have but little difficulty in recognising most of the principal stars in the district

o

occupied by Pegasus, from the directions previously given; the following lines will, however, still further assist him :—

> " And on, from where the pinioned maid
> Her cruel fate attends,
> Wide o'er the heavens his fabled form
> Winged Pegasus extends.
> From Alpherat down to Markab's beams
> Let a cross-line be sent,
> Then will four stars upon the Horse
> A spacious square present."

PERSEUS.

Perseus is principally a northern constellation, and for the most part circumpolar. It is one of the forty-eight asterisms of the ancients, and is situated in a very conspicuous part of the Milky Way, directly north of the Pleiades. Its chief stars are Alpha Persei, sometimes called Mirfak, and Beta Persei, or Algol. The latter is one of the most remarkable of the variable, or periodic, stars. The variation in lustre of this star was first noticed by Montanari in the seventeenth century, but its periodicity was first accurately determined in 1782 by Goodricke. Algol varies in magnitude from the second to the fourth in about three and a half hours, and back again to the second in the same interval of time. It continues at its greatest lustre during the remainder of its period, which has been ascertained to be about two days, twenty hours, and forty-nine minutes. Perseus contains several interesting stellar objects, one of which in his right hand we give as an illustration. This beautiful cluster is scarcely visible to the naked eye, even as a single star, but when viewed through a good telescope it exhibits a brilliant mass of stars, varying from the seventh to the fifteenth magnitudes. While gazing on this superb telescopic object we can well realise the poet's description of—

> " Some sequestered star
> That rolls in its Creator's beams afar,
> Unseen by man ; till telescopic eye,
> Sounding the blue abysses of the sky,
> Draws forth its hidden beauty into light,
> And adds a jewel to the crown of night."

STAR CLUSTER IN PERSEUS.

In this beautiful cluster, the central group resembles a coronet, or rather an ellipse of small stars. The comparatively bright star to the right in the diagram is of the seventh magnitude. Sir William Herschel considered that this cluster, and another which follows it closely, belong to the Milky Way, in which they are situated. These two clusters are perfectly disconnected from each other, although the outlying stars in each can be brought into the field of view of a telescope at the same time. On very clear nights in winter they form, when taken together, one of the most interesting telescopic objects in the heavens. The author recollects the great delight he felt when he was first shown these clusters, at a time when such objects were novel to him; and although thirty years of professional experience have passed away since then, the impression on his memory of this first view of these gorgeous groups has never been effaced.

Perseus is situated in the Milky Way, south of Cassiopeia and Camelopardalis, east of Triangulum and Andromeda, north of Taurus, and west of Auriga and Camelopardalis. It contains about sixty stars visible without the aid of a telescope.

SAGITTA.

Although one of the most insignificant of the constellations of the northern hemisphere, Sagitta, or the Arrow, is included among the forty-eight asterisms of the ancients. It is represented in celestial maps with the point of the arrow towards the east, occupying a very narrow portion of the heavens between the bill of the Swan and Aquila. One of the branches of the Milky Way passes through Sagitta. On the shaft of the arrow a rich but compressed cluster, 71 (Messier) Sagittæ, may be observed with a good telescope. Under ordinary circumstances, it appears as a nebula with a very feeble light, to which class of objects it was supposed to belong, till Sir William Herschel, in 1783, resolved it into a cluster of stars. Flamsteed, in his *Historia Celestis*, recorded the positions of eighteen stars in Sagitta, nearly all of which are discernible with the unassisted eye.

SERPENS.

Serpens is one of the forty-eight ancient constellated groups, and extends over a considerable portion of the sky; the head, which is under Corona Borealis, is well marked by several stars of the third magnitude; the body winds through Ophiuchus, and the tail reaches the Milky Way near Aquila. Its principal star, Alpha Serpentis, of the second and a half magnitude, was known to the ancients by the name of Unukalkay, and by the astrologers of the middle ages as Cor Serpentis, or the Heart of the Serpent. This star is of a pale yellow colour, and is closely followed by a very small telescopic object, first noticed by Sir William Herschel with his twenty-foot reflector. In the catalogues of Ptolemy and Copernicus this constellation is made to consist of eighteen stars, all of which were clearly visible to the naked eye; Flamsteed increased this number to sixty-four, while the atlas of Bode contains one hundred and eighty-seven. Most of the principal stars in Serpens are situated in or near the head; Alpha is, however, in the fore-part of the body, a short distance below the head. The reader can easily identify the last-

163

mentioned star in the June diagram of the sky, looking south, by drawing a straight line from Alpha Aquilæ, or Altair, east of the meridian, through Alpha Ophiuchi, or Ras Alague, on the meridian, to a corresponding distance west of the meridian, when it will pass through, or very near, Alpha Serpentis. Or, if we have recourse, as on other occasions, to the rhymester, the upper part of the body of the Serpent may be clearly pointed out by reference to the same stars :—

> " To strike th' insidious Serpent's heart,
> A line from Altair wield,
> From thence below Ras Alague,
> Across th' Arabian Field ;
> And when as far again you've reach'd,
> As those two stars may be,
> The middle one of three fair gems,
> Serpentis Cor you'll see."

TRIANGULUM.

Triangulum, the Triangle, is also one of the ancient forty-eight constellated groups. It occupies, however, a very limited space south-east of Andromeda, north of Aries, and west of Perseus. Hevelius formed an additional triangle, which he named *Triangulum Minor*, by appropriating a few unused stars between it and Aries. At the present time, however, this creation of Hevelius has been discontinued. Alpha Trianguli, the principal star, is about the third magnitude, accompanied by a very faint telescopic companion; the remaining stars visible to the naked eye vary from the fourth to the sixth magnitudes.

URSA MAJOR.

Ursa Major, the Great Bear, is perhaps the most universally known of all the constellations in the northern hemisphere. The familiar form of its seven principal stars, being always visible in the latitude of Great Britain, naturally contributes to make it so. One of the first lessons on the configurations of the stars, which we should advise the young student to make, would be the identification by name of each of the stars composing this remarkable group; for

> " He who would scan the figured skies,
> Its brightest gems to tell,
> Must first direct his mind's eye north,
> And learn the Bear's stars well."

Ursa Major extends over a considerable portion of the northern sky. It has Draco and Camelopardalis on the north, Canes Venatici on the east, Leo Minor on the south, and Lynx and Camelopardalis on the west. Excepting Draco, it happens that in the constellations directly surrounding Ursa Major the sky is more than usually bare, which tends to show out its seven principal stars all the brighter by the contrast. The different members of this group are nearly of the same magnitude, six of them being about the second. The seventh, Delta, is sensibly smaller than the others, and it has been suspected to be variable in lustre at intervals of long period. At the present time, it is ranked only of the third and a half magnitude, while in the days of Tycho Brahé, Delta was recorded as of the second, and in the seventeenth and eighteenth centuries of the third and fourth respectively. A mere glance at the group shows distinctly that it can bear no comparison in lustre with its six companions.

This popular stellar group has been for a long period associated in name with several terrestrial objects, to which it has been assumed to bear some resemblance. Among them we have the appellations of Charles's Wain, the Plough, David's Car, the Bier of Lazarus, the Dipper, etc. Charles's Wain is the name by which it is most known in this country. In this instance, the four stars, Alpha, Beta, Gamma, and Delta, are termed the wheels, the two first being the fore wheels, while the remaining three represent the shaft of the waggon. These last in the Plough would represent the handle, and also of the Dipper. One zealous writer, Kircher, claims the four stars in the quadrilateral as the Bier of Lazarus, the three in Ursa's tail representing Mary, Martha, and Mary Magdalene. The poet Schiller has, in imagination, seen the ship of St. Peter typified by this group. With regard to the resemblance of these stars to the form of a Bear, no very great stretch of poetical fancy is required to make out a rough form of that animal ; for, in addition to the body and tail, indicated by the principal stars, there are others of the third and fourth magnitude that fix the outline of this imaginary Bear with an astonishing degree of precision.

The two most westerly, or advanced, of the seven stars are universally known as the Pointers. These two stars, Alpha and Beta Ursæ Majoris, have also their Arabic names, Dubhe and Merak, and are, perhaps, the most popular of the seven, because, by drawing a line from the more southerly of the two through the other, the Pole-star is always found without difficulty. There is not much fear that this operation would fail, even with the youngest observer.

Mizar, the central object of the three in the tail, is a splendid double star, the companion being a bluish telescopic star of the eighth magnitude. Alcor, of the fifth magnitude, is some distance from Mizar, when seen through a telescope, but with the naked eye the two have the appearance of a double star. The last star in the tail has the Arabic designation Alkaid, or Benetnasch, or the superintendent of the mourners, supposed to be following the fanciful form of a bier. The following statement in a tabular form gives the astronomical name, with its Arabic designation, of each of the stars in Charles's Wain ;—

Alpha (a)	Ursæ Majoris or		Dubhe,
Beta (β)	,,	,,	Merak,
Gamma (γ)	,,	,,	Phecda,
Delta (δ)	,,	,,	Megrez,
Epsilon (ε)	,,	,,	Alioth,
Zeta (ζ)	,,	,,	Mizar,
Eta (η)	,,	,,	Alkaid, or Benetnasch.

URSA MINOR.

Ursa Minor, the Lesser Bear, is very easily distinguished, as its contour, marked out by seven stars, as in Ursa Major, can be traced with ease from Polaris in the tail to Kocab and Gamma Ursæ Minoris in the shoulders. The curvature of the tail is, however, in a contrary direction to that of Ursa Major. This small constellation, which contains the pole of the heavens, is surrounded by Draco, Camelopardalis, Cassiopeia, and Perseus. Its principal star, Polaris, of the second magnitude, is one of the most important objects in the heavens.

Polaris is a yellowish star, accompanied by a faint, but not close, companion of the ninth magnitude. Several observers have made measures of their distance and angle of position at different epochs, to determine whether the two stars are physically, or only optically, connected. The evidence deduced from the observations seems to show that, though they are telescopic companions, yet they are probably separated from each other by an almost infinite distance. In the northern hemisphere, Polaris is of great advantage to the astronomer and mariner, being used frequently for determining the latitude of places. Being of a good magnitude, it is visible through a telescope on the meridian twice in twenty-four hours; once above the pole, and once below the pole. As the angular elevation of the celestial pole above the horizon is always the same as the latitude of the place of observation, the mean or average of the two measurements of meridian altitudes above and below the pole determines the latitude. The same can be found from any circumpolar star bright enough to be seen during the daytime; but, practically, Polaris is the best star for the purpose, as the small angular distance between the two positions is favourable for several reasons. This star is also observed regularly in most standard observatories for the determination of the azimuthal error of the transit-instrument, or the deviation of the telescope from the true meridian. Polaris is becoming year by year more truly the Pole-star, as it will be gradually approaching the pole for the next 200 years or more. It will then begin to recede, continuing to do so for about 12,000 years, when the pole of the heavens will be somewhere in the constellation Lyra. In 12,000 years more it will be again in the same position as at present. In the interval of time between these distant epochs, Polaris will therefore cease for ages to bear the name of, or be useful in any way as, the Pole-star.

The use of this star in navigation is supposed to have been first recommended by the Greek astronomer Thales, and was very anciently known by the name of Phœnice. The poet Dryden has described the infancy of navigation, as practised by the adventurous seamen of Phœnicia:—

> " Rude as their ships were navigated then,
> No useful compass, or meridian known ;
> Coasting they kept the land within their ken,
> And knew no north but when the pole-star shone."

The second bright star in Ursa Minor is Beta, or, as named by the Arabs, Kocab. It is of about the second magnitude, and is situated in the left shoulder of the Lesser Bear, Polaris being at the farthermost end of the tail. The third star, Gamma, is of the third magnitude. Kocab and Gamma Ursa Minoris are frequently alluded to as the Guards of the Pole. The intermediate stars between Beta and Gamma at one extremity and Polaris at the other are much smaller ; but on moonless nights they are quite distinct, and easily identified. Perhaps the following lines may help the observer to find the stars in Ursa Minor :—

> " Kocab, one bright, and two faint stars,
> Grace Lesser Ursa's side,
> In oblong square ; trace her bent tail,
> And to the Pole you'll glide."

VULPECULA ET ANSER.

Vulpecula et Anser, the Fox and the Goose, between Aquila and Cygnus, is one of the modern constellations introduced by Hevelius in the seventeenth century. " I wished," said he, " to place a fox with a goose in the space of a sky well fitted to it ; because such an animal is very cunning, voracious, and fierce. Aquila and Vultur are of the same nature, rapacious and greedy." The small stars below Albiero belong to Vulpecula. Hevelius registered twenty-seven stars in this small asterism, and Bode as many as one hundred and twenty-six.

In addition to the preceding constellations, there are several others north of the Zodiac, including Mons Mænalus, Musca Borealis—the Northern Fly, Quadrans Muralis—the Mural Quadrant, Scutum Sobieski—the Shield of Sobieski, Tarandus—the Reindeer, Taurus Poniatowski—the Bull of Poniatowski, and some even more insignificant than these. Not one of these asterisms contains a single star likely to attract attention.

CONSTELLATIONS SOUTH OF THE ZODIAC.

"How bright the starry diamonds shine! The kings
Of eastern climes enjoy'd a pride to be
Compared with them,—they deck'd their robes with stars.
See how they glitter through the void! They form
Night's richest dress; they on her sable robe
Like gems of richest lustre sparkle, spread
So in extent, no nation so remote
But sees their beauty."

THE principal constellations south of the zodiac, many of which are visible above the horizon of London, are Ara, Argo Navis, Canis Major, Canis Minor, Centaurus, Cetus, Columba Noachi, Corona Australis, Corvus, Crater, Crux Australis, Dorado, Eridanus, Grus, Hydra, Hydrus, Indus, Lepus, Lupus, Monoceros, Musca Australis, Orion, Pavo, Phœnix, Piscis Australis, Piscis Volans, Robur Caroli, Sextans, Triangulum Australis, and Toucana. There are many others, composed principally of small stars, introduced by the astronomers of the seventeenth and eighteenth centuries to fill up certain vacancies between the larger constellations. In the following notes we have confined ourselves to those which contain well-known groupings of stars, or single stars of large magnitude, having special interest.

ARGO NAVIS.

The great constellation Argo Navis, whether we regard it on account of its relative position in the southern heavens, or for the numerous stars of superior magnitude contained in it, holds a high place among the principal asterisms either north or south of the zodiac. It includes within its boundaries two stars of the first magnitude, five of the second, and ten of the third, with the usual number of fourth, fifth, and sixth magnitudes, all visible to the unassisted eye. Argo Navis, when on the upper meridian, between the south pole and the zenith, has Hydra and Monoceros on the north, Robur Caroli on the east, several small constellations on the south, and Canis Major and Columba Noachi on the west. The ship Argo has been divided by astronomers into four compartments, named respectively the hull, the keel, the stern or poop, and the sail. A small part of the ship's poop is the only portion of this constellation which rises above the horizon of Great Britain. Canopus, the principal star, is near the keel or rudder of the ship, and is the most westerly large object in Argo Navis when between the pole and the zenith.

There is one important object in Argo Navis, far more interesting to the astronomer than the brilliancy of its individual members, known as "the great nebula in Argo." Since Sir John Herschel made a detailed drawing of it at Feldhausen from observations made from 1834 to 1837, several most remarkable changes have apparently taken place, not only in the general aspect of the nebula, but also in the relative positions of the neighbouring stars. According to that illustrious astronomer, whose observations in South Africa will go down to posterity as a lasting monument of his zeal and devotion to stellar astronomy, this wonderful nebula is spread over an area of at least a square degree, and is one of those rich and brilliant masses of minute stars, a succession of which contrasts most curiously with dark adjacent spaces, distinguished by old navigators by the name of coal-sacks, forming the most attractive portion of the Milky Way, near Centaurus and the Southern Cross. In the midst of the vast stratum of stars or nebulous matter composing this nebula in Argo Navis, the remarkable variable star Eta Argûs is situated. When Dr. Halley observed Eta Argûs in the year 1677, it was of the fourth magnitude. When Sir John Herschel first saw it in 1834 it appeared between the first and second magnitudes, in which condition it remained without any apparent change up to November, 1837. When compared with other stars, it was always considered to be superior in brightness to Beta and Gamma Crucis, Beta Argûs, and Pollux, and inferior to Alpha Crucis, Antares, Spica, and Aldebaran.

"It was on the 16th December, 1837," remarks Sir John Herschel, "that resuming the photometrical comparisons in question, in which, according to regular practice, the brightest stars in sight, in whatever part of the heavens, were first noticed, and arranged on a list, my astonishment was excited by the appearance of a new candidate for distinction among the very brightest stars of the first magnitude, in a part of the heavens with which, being perfectly familiar, I was certain that no such brilliant object had before been seen. After a momentary hesitation, the natural consequence of a phenomenon so utterly unexpected, and referring to a map for its configurations with the other conspicuous stars in the neighbourhood, I became satisfied of its identity with my old acquaintance Eta Argûs. Its light was, however, nearly tripled. While yet low it equalled Rigel, and when it had attained some altitude was decidedly greater. It was far superior to Achernar. Fomalhaut and Alpha Gruis were at the time not quite so high, and Alpha Crucis much lower, but all were fine and clear, and Eta Argûs would not bear to be lowered to their standard. It very decidedly surpassed Procyon, which was about the same altitude, and was far superior to Aldebaran. It exceeded Betelgeuse; and the only star, Sirius and Canopus excepted, which could at all be compared

P

with it was Rigel, which, as I have stated already, it somewhat surpassed. From this time its light continued to increase. On the 28th December it was far superior to Rigel, and could only be compared with Alpha Centauri, which it equalled, having the advantage of altitude, but fell somewhat short of it as the altitudes approached equality. The maximum of brightness seems to have been obtained about the 2nd January, 1838, on which night, both stars being high, and the sky clear and pure, it was judged to be very nearly indeed matched with Alpha Centauri, sometimes the one, sometimes the other, being judged brighter, but on the whole Alpha was considered to have some little superiority. After this the light began to fade."

The gradual diminution of the intrinsic light of Eta Argûs, observed by Sir John Herschel, was only of a temporary nature, for in 1843 it again appeared brighter than Canopus, and at one time it even approached Sirius in brilliancy. The attention of Sir John Herschel, who had returned to England, was first drawn to the renewed splendour of Eta Argûs by the Rev. W. S. Mackay, of the General Assembly's Mission, Calcutta, and subsequently by Mr. (now Sir Thomas) Maclear, the Government astronomer at the Cape of Good Hope.

In a letter dated September 17, 1844, Mr. Maclear, referring to his observations of this star, says "that the changes of Eta Argûs are curious, for last April twelve months it seemed almost equal to Sirius. Now, the light of the star is stationary, and scarcely so bright as Canopus." In 1845, although Eta Argûs appeared of the first magnitude, it evidently at that time was declining gradually. This diminution continued from year to year until 1863, when it shone no brighter than a star of the sixth magnitude. It has remained in this condition to the present time, thus no longer pointing out to the naked eye, without an effort, the position of the great nebula in Argo. When at its greatest lustre, Eta Argûs materially interfered with the light of the nebula in the vicinity of the star, and obliterated completely some of the fainter portions of it.

Some recent observations and drawings made by Mr. Abbott, at Hobart Town, Tasmania, of the relative positions of the principal stars in this great nebula, have again directed the attention of astronomers to this wonderful object. By comparing Mr. Abbott's drawings with that made in 1837 by Sir John Herschel, it is at once perceived that the position of the star Eta Argûs is very different, in Mr. Abbott's delineation of the nebula, from that recorded by Sir John Herschel. In the intermediate time, the form of the whole nebula appears to have undergone a complete change. Moreover, stars which were observed in 1834–37 seem to have vanished altogether, while others not perceived in those years have apparently come into existence. Sir John Herschel, writing on this remarkable change in the form of the nebula, remarks that " there is no phenomenon in nebulous or sidereal astronomy that has yet turned up, presenting anything like the interest of this, or calculated to raise so many and such momentous points for inquiry and speculation. The question here is not of minute variations in subordinate features, which may or may not be attributable to differences of optical power in the instruments used by different observers, but of a total change of form and character—a complete subversion of all the greatest and most striking features—accompanied with an amount of relative movement between the star and the nebula, and of the brighter portions of the latter *inter se*, which reminds us more of the capricious changes of form and place in a cloud drifted by the wind, than of anything heretofore witnessed in the sidereal heavens."

Attention being now directed to these remarkable changes, it is expected that a series of drawings will be made, at stated intervals, by some of the astronomers at our southern observatories, which will probably increase our knowledge of the physical composition of this nebula, as well as settle the interesting question as to its supposed transitional state. As seen by the naked eye, it has been stated that the beautiful soft white light generated by the nebulous matter, may be produced by the comparatively superior magnitude of the small stars within its boundaries. According to Mr. Abbott, its contrasting brightness with the dark neighbouring sky is very distinctly marked; for "on a clear, fine night the object gives out fully twice as much light as that of the great nebula, the Nubecula Major, and about three times as much as the Nubecula Minor, irrespective of size. In the twilight it appears as soon as a star of the second or third magnitude, the light being white and more diffuse, very like a small white woolly cloud on a blue sky, seen in sunlight." In a drawing of this nebula, made in India in 1868 by Lieut. Herschel, the apparent change is not so great as that depicted by Mr. Abbott.

CANIS MAJOR.

Canis Major, the Great Dog, is below Orion—Sirius, the principal star, being reputed to be one of the hounds of that noted celestial warrior. Besides Sirius, Canis Major contains several stars of the second and third magnitude, most of which, in the latitude of London, are clearly visible below Sirius, near the horizon. From the brilliancy of this fine object, which is situated in the mouth of the Dog, it is scarcely possible not to identify it almost at a glance. By reference, however, to other large stars, there are several ways by which it can be pointed out. For example, a line drawn from the Pleiades through the three stars forming Orion's belt leads directly to it; and, as we have already mentioned, it forms with Betelgeuse and Procyon a fair equilateral triangle. Sirius is a perfectly white star, though it has been asserted that some centuries ago it had a reddish appearance. In the time of Ptolemy, who flourished in the reigns of the Roman Emperors Adrian and Antoninus, Sirius was recorded of a red colour. That prince of ancient astronomers, in his celebrated catalogue of the fixed stars, contained in the *Almagest*, has put down the following stars as being of a fiery red colour: Arcturus, Antares, Aldebaran, Betelgeuse, Sirius, and Pollux. Of these, Arcturus and Antares have still a fiery red appearance, Aldebaran is of a rose tint, Betelgeuse and Pollux have an orange tinge, while Sirius is of a brilliant white. Alexander von Humboldt is of opinion that, taking for granted that the colour of Sirius had changed at some time from red to white, a great physical revolution must have taken place on the surface, or in the photosphere of this fixed star, " before the process could have been disturbed, by means of which the less refrangible rays had obtained the preponderance through the abstraction or absorption of other complementary rays, either in the photosphere of the star itself, or in the moving cosmical clouds by which it is surrounded." It is to be regretted that, in the interval between the time of Ptolemy and the present day, no reference appears to have been made in history or poetry to this alleged remarkable change in the colour of Sirius. In the time of Tycho Brahé, however, we may reasonably infer that this star was white as at present, from the following circumstance:— In the year 1572, a celebrated temporary star suddenly appeared in Cassiopeia. During its continuance, Tycho observed it to change in colour in a short period from a dazzling white to a ruddy hue, comparing it to the colour of Mars and Aldebaran. If at that time Sirius had been as red as in the days of Ptolemy, it would more naturally have been one of the stars of comparison, instead of, or as well as, Aldebaran, on account of its superior magnitude. It has, therefore, been concluded that the colour of this brilliant star is the same now as in 1572, and if any change has taken place it must have been antecedent to that date.

The brilliant appearance of Sirius among the stars must have attracted, in all ages, not only the attention of astronomers, but also of every person who occasionally gives a passing thought to the wonders of the universe above and around him. When this star was made to enter the field of view of Sir William Herschel's great forty-foot reflecting telescope, the glow of light, before it became visible to the observer, gave the appearance of the approach of sunrise; and when the star was fairly in the centre, the glare was always so great that it was scarcely possible to keep the eye directed to it without inconvenience, if not actual pain. Even in refracting telescopes with large object-glasses, the image of this star is exceedingly bright, though not equal in intensity to that produced from reflection from such large polished metallic surfaces as those contained in the telescopes of Sir William Herschel or the Earl of Rosse. From the photometric observations of Sir John Herschel, the intensity of the light of Sirius has been found equal to 324 stars of the sixth magnitude.

A most remarkable series of researches has been made on the apparent irregular motion of Sirius in the heavens. Here it will be necessary to mention that, though the stars are termed " fixed stars," yet they are only comparatively so with respect to the planets of our solar system. For it has been found that almost every star has its own peculiar motion, small indeed as it appears to us, but still sufficiently large to be detected after many years' observations with standard meridional instruments. Now it has been discovered, not very long ago, that this peculiar movement of Sirius is not regular like the rest of the stars, but that it is greater or less in different years, or in different series of years. The cause appeared inexplicable, till two astronomers undertook independently to investigate the subject by the application of the highest branches of mathematical analysis. It was soon announced that these irregular movements of Sirius could only be accounted for by supposing it to be affected by the attraction of some neighbouring body of sufficient magnitude. To detect this body, or satellite, was looked upon as a hopeless task, as it could only be expected to be found within the bright rays of the star. However, Mr. Alvan Clark, of Boston, United States, with a powerful telescope of his own make, noticed a very small object on January 31st, 1862, while viewing Sirius under very favourable circumstances. After this, several astronomers in Europe and America have not only seen this small object, but have also succeeded in measuring its distance and angle of position in relation to the large star. By comparing the observations made in different years with the results deduced from theory, the agreement has been found to be very close. It is therefore now believed that a planet or satellite has been discovered which is evidently a member of the Sirius system, and that it is of sufficient magnitude to influence the movements of the central body.

From the heliacal rising of Sirius the ancients reckoned the *dies caniculares*, or dog-days. It does not, however, require much acquaintance with astronomy to know that the commencement of this season can really have but little connection with the rising of Sirius, for in different latitudes the heliacal rising of that star varies considerably. There are other astronomical reasons which tend to show that Sirius is not guilty of the many evils attributed to his rising with the Sun. As far as this country is concerned, the dog-days of the present generation can have no reference whatever to the rising of Sirius, for almanack-makers always include that usually warm period between July 3rd and August 11th, while Sirius rises heliacally on August 25th, or thirteen days after the conclusion of the dog-days. The ancients, however, believed faithfully in the reputed unfavourable influences of Sirius on various kinds of diseases. Theon Alexandrinus, an astronomer in the olden time, left several precepts, among which was one "to find the exact time of the Dog-star's rising with the Sun ; twenty days before which, and twenty days after, included the period of extreme heat, hydrophobia, and other evils."

CANIS MINOR.

Canis Minor, the Lesser Dog, is south of Gemini, west of Hydra, and north-east of Canis Major, the Milky Way passing between it and the last-mentioned constellation. Canis Minor has always been regarded with great popular interest. In ancient times Procyon, the principal star in Canis Minor, was called the Precursor Dog, from its appearing in the morning dawn shortly before Sirius. Among the Arabians it was recognised not only as the forerunner of the Dog-star, but as the bright star of Syria, as well as of the Lesser Dog. This interest of former ages evidently descended to the astrologers of later times, one of whom, Leonard Digges, has remarked, "What meteoroscoper, yea, who learned in matters astronomical, noteth not the great effects at the rising of the starre called the Litel Dogge ?" Procyon is of the first magnitude, and is situated in the centre of the body of the animal. In the diagrams of the sky of the northern hemisphere, it is easily found, being below Castor and Pollux. It is also one of the stars forming the triangle with Sirius and Betelgeuse already alluded to. Another way of pointing out

Procyon is by drawing a line from the three stars in Orion's belt to Sirius, then one perpendicularly raised over the latter star will pass through Procyon towards the north. This alignment has been put into rhyme thus :—

> "Orion's belt from Taurus' eye
> Leads down to Sirius bright ;
> His spreading shoulders guide you east,
> 'Bove Procyon's pleasing light."

Canis Minor contains but few prominent stars. Beta Canis Minoris, to the west of Procyon, is of the third magnitude ; the remainder are all of inferior lustre. Ptolemy recorded the positions of only two stars in Canis Minor, Tycho Brahé and Kepler five. Bode's Atlas contains fifty-five, the majority of which are, however, telescopic.

CENTAURUS.

The name of Centaurus, one of the ancient asterisms, is supposed to have been derived from the sons of Ixion, who were fabulously represented as half men and half horses. This constellation is partly visible above the south horizon of London, but its principal stars are only seen in more southern latitudes. It is bounded on the north by Hydra, on the east by Lupus, on the south by Crux Australis, and on the west by Robur Caroli, or King Charles's Oak. Centaurus contains more than an average number of bright stars, two of the first magnitude, one of the second, six of the third, and a good number of the fourth. They are all included in the most brilliant portion of the southern sky. Alpha and Beta Centauri, in the fore-feet of the Centaur, contribute in no small degree to this brilliancy. Alpha Centauri, in the right fore-foot, is a most celebrated double and binary star, and is one of the nearest to our solar system. In the infancy of telescopes, when their optical power was small, this object was only seen as a single star ; for Richer observed it at Cayenne in 1673, and Halley at St. Helena in 1677, without mentioning its duplicity. A scientific French traveller, M. Louis Feuillée, was the first person who saw the star divided into two. In the journal of his observations made in South America, he states that, being at Concepcion, in Chili, in July, 1709, " I observed, with a telescope of eighteen feet focal length, the star of the first magnitude which is in the northern fore-foot of Centaurus ; I found this star composed of two, of which one is of the third, and the other of the fourth magnitude. That of the fourth magnitude is the more westerly, and their distance is equal to a diameter of that star." It is recorded in the *Philosophical Transactions* for 1749, that M. La Condamine also saw this star double whilst on a scientific expedition to Peru, for the purpose of measuring the exact value of an equatorial arc of the meridian. La Condamine states that Alpha Centauri rivals Capella in splendour and magnitude ; and, with a small telescope of three feet focal length, it appeared double, consisting of two stars, of which the lesser seemed to emerge from the greater. With the improved achromatic astronomical telescopes of the present day, the two stars appear separated by a very appreciable angular distance. Mr. Dunlop, at Paramatta, New South Wales, found the distance in 1825 to be equal to 25″. Yearly observations showed that the space between the two stars was diminishing at the rate of half a second per year. At present it is no more than about 7″. These stated comparisons show that the two objects are physically connected, and that they form one of the most interesting of the binary stars.

Extensive series of observations have been made at the Royal Observatory, Cape of Good Hope, by two successive astronomers, Messrs. Henderson and Maclear, for the purpose of determining the annual parallax of Alpha Centauri, from which has been deduced its approximate distance from the Earth. It was found, without much probability of error, that the amount of the apparent displacement of this star in the heavens, produced by parallax when viewed from opposite parts of the Earth's orbit, is nearly one second of arc. From this value it has been determined that the distance of Alpha Centauri, probably the nearest fixed star to our solar system, is more than two hundred thousand times greater than that of the Earth from the Sun. We know with tolerable certainty that the latter distance is about ninety-two millions of miles; but to ascertain that of the star, it is necessary to multiply these numbers by 200,000. It is quite impossible for ordinary minds to grasp such immense distances in figures; but if we were to compute the time required for light to pass from the star to the Earth, or for an express train, with no stoppages, to traverse that vast interval of space, some slight idea may be gathered of the enormous distance of this nearest of the stars. From the most recent investigations, it has been found that light travels at the rate of about 185,000 miles a second, passing through the intermediate space between the Earth and Sun in 8^m 18^s. By a simple computation, any one may be able to find that, even at this tremendous velocity, light will take upwards of three years to reach the Earth after it has been emitted by this star. An express train, travelling with a speed of sixty miles an hour, would not arrive at the end of such a journey before thirty-five millions of years had passed away.

The intrinsic light cast upon the Earth by the largest of the fixed stars, bears no comparison with that of the full Moon. Several most interesting photometric experiments were made by Sir John Herschel at Feldhausen, near Cape Town, on the relative light of the principal stars, especially on that of Alpha Centauri, in comparison with the light of the full Moon. Sir John Herschel found that the full lunar light exceeded that of Alpha Centauri in the proportion of 27,408 to 1. Dr. Wollaston had previously found that the proportion of the Sun's light to that of the full Moon equalled 801,072 to 1. Hence the light sent to us by the Sun is twenty-two thousand million times greater than that reaching the Earth from the star. But if we were to assume that our Sun and Alpha Centauri were equally distant from us, the intensity of the star's light would be nearly two and a half times more brilliant than that proceeding from our central luminary. The results of these observations of Sir John Herschel and Dr. Wollaston cannot fail to give us some conception of the immense magnitude and intrinsic brilliancy of the thousands of stellar suns visible to us one-very cloudless night.

CETUS.

Cetus, the Whale, is south of Aries and Pisces, west of Eridanus, Orion, and Taurus, east of Aquarius, and north of Sculptor and other small constellations. Cetus is very extensive, and is one of the old standard forty-eight asterisms. It contains three hundred stars of sufficient magnitude to be included in Bode's Atlas. Alpha Ceti, or Menkar, and Beta Ceti, or Diphda, are the principal stars. They are, however, widely separated, Menkar being near the eastern portion of the constellation, and Diphda the western. The index-maps will point out the relative positions of these objects; but, by star-alignment, a line drawn from Pollux to Aldebaran, and then carried forward nearly as far again, will pass close to Menkar, in the head of the Whale. Or, according to the rhymester:—

> " To know the bright star in the Whale,
>   The lower jaw which decks ;
> From fair Capella send a glance
>   Through Pleiad's beauteous specks ;
> And bear in mind this cluster fine,
>   So admirably seen,
> From Cetus' head to the Charioteer,
>   Lies just half-way between."

A very remarkable variable star is to be found in this constellation, known by the name of Mira Ceti. The variation in lustre of this star was first noticed by David Fabricius, in 1596. It retains its maximum brightness during fourteen days, and is then of the second magnitude. Its light afterwards gradually decreases for about three months, when it becomes not only invisible to the naked eye, but also when looked for with the largest telescopes. It remains invisible during five months, then reappears as a minute telescopic object, and afterwards increases gradually for three months, when it again attains its maximum splendour. The time of its period from maximum to maximum is about 331 days. The greatest lustre of this curious star has been found to be not always the same; it is usually equal to the second magnitude, but, occasionally, it has been recorded that at its maximum it has appeared only of the fourth magnitude. In 1799, according to Humboldt, its light shone with an intensity nearly equal to that of stars of the first class, in fact scarcely inferior to Aldebaran. This object is one of the most interesting of the variable stars, as may be inferred from the appellation by which it is distinguished, Mira, or the wonderful star.

COLUMBA NOACHI.

This constellation, instituted by Royer in 1679, is of small extent, but it contains a few bright stars. Alpha Columbæ is of the second magnitude. This star can be seen, in the latitude of London, very near the south horizon when on the meridian. It is visible directly below Lepus. Columba lies west of Argo Navis and Canis Major. In the southern hemisphere, its principal stars are therefore very favourably situated, passing the meridian in or near the zenith. This small constellation contains about twenty-six stars clearly visible without telescopic aid.

CORVUS.

Corvus, the Crow, is also a small constellation. The bird is supposed to rest on the body of Hydra, a part of which is included in this asterism. Corvus is one of the ancient constellated groups, and it can readily be made out in the heavens by its four stars of about the third magnitude in a district not very rich in large stars. It contains about ten stars visible to the unassisted eye. A long line drawn from Vega through Spica carried farther on about fifteen degrees, passes through the four stars by which Corvus is generally distinguished. Or—

"Mark in the space along the sky,
Where Hydra's volumes are,
And 'twixt the Cup and Virgin's spike,
You'll find the Raven's square."

CRATER.

Like Corvus, Crater, the Cup, is situated on the back of Hydra, its form being easily made out by a number of stars of about the fourth magnitude. Crater lies east and north of Hydra, west of Corvus, and south of Virgo. It is one of the old constellations. Ptolemy's catalogue contains seven stars in Crater.

CRUX AUSTRALIS.

Crux Australis, or the Southern Cross, although the subject of universal attraction to all star-gazers south of the equator, is a very small constellation. Taken by itself away from the brilliancy of the neighbouring sky, much of its attraction would disappear. The upper and lower stars being of similar right ascension, they are always on the meridian about the same time, and consequently serve, like the pointers in Ursa Major, to indicate the approximate position of the south pole, which is distant about 27° 38′ from the largest and nearest star in the Cross. At and south of the equator this constellation can be well seen, while in the latitude of 34° S. it never sets below the horizon. It is, therefore, always visible at the Cape of Good Hope, Australia, etc., just in the same manner as the Great Bear never sets at London. There was a time, nearly five thousand years ago, when the Southern Cross was visible even from the shores of the Baltic, but at too low an elevation to be very distinctly seen. At the present time it is annually receding from the south pole by a small but regular quantity, and the day will probably again come, at a distant age, when it will reappear above the horizon of Europe.

In a former chapter we have alluded to the interesting first impressions which the sight of this well-known group of stars makes on the mind of the traveller on passing from the northern to the southern hemisphere. Those recorded by MM. Von Spix and Karl Von Martius, in their account of their scientific travels in Brazil, in 1817–1820, give a very fair specimen of the feelings experienced on these occasions. It is related by them that " on the 15th of June, in latitude 14° S., we beheld for the first time that glorious constellation of the southern heavens, the Cross, which is to navigators a token of peace, and, according to its position, indicates the hours of the night. We had long wished for this constellation as a guide to the other hemisphere; we therefore felt inexpressible pleasure when we perceived it in the resplendent firmament. We all contemplated it with feelings of profound devotion, as a type of our salvation; but the mind was especially elevated at the sight of it, by the reflection that even into the region which this beautiful constellation illumines, under the significant name of the Cross, the European has carried the noblest attributes of Christianity, and, impelled by the most exalted feelings, endeavours to spread them more and more extensively in the remotest regions." The scientific Humboldt has expressed his thoughts in almost similar terms. Referring to his first view

THE CONSTELLATION OF THE SOUTHERN CROSS.

of the constellation, he observes that, " We saw distinctly, for the first time, the *Cross of the South*, on the night of the fourth and fifth of July, in the sixteenth degree of latitude; it was strongly inclined, and appeared from time to time between the clouds, the centre of which, furrowed by uncondensed lightnings, reflected a silver light. The pleasure felt on discovering the Southern Cross was warmly shared by such of the crew as had lived in the colonies. In the solitude of the seas we hail a star as a friend, from whom we have been long separated. Among the Portuguese and the Spaniards, peculiar motives seem to increase this feeling; a religious sentiment attaches them to a constellation, the form of which recalls the sign of the faith planted by their ancestors in the deserts of the New World." Poets, as well as travellers, have also exercised their imaginative powers on so inviting a theme as that presented to them by the congregation of so much stellar beauty in the limited region in and around the neighbourhood of the Southern Cross. Even Dante appears to have had some dim knowledge of the existence of these stars, although he flourished long before the date of Vasco da Gama's famous expedition to the Indies. In his *Divina Commedia* the following passage possibly alludes to this constellation:—

> " To the right hand I turned, and fix'd my mind
>   On the other pole attentive, where I saw
> Four stars ne'er seen before, save by the ken
>   Of our first parents. Heaven of their rays
> Seemed joyous. O thou northern site ! bereft
>   Indeed, and widow'd since of these deprived."

On first thoughts, the above extract may appear to be simply a poetic fancy; but a little consideration will show that a vague tradition may have reached Europe, anterior to the time of Dante, of the existence of other bright celestial objects besides those visible in the northern heavens. It is a fact, not to be disputed, that the Arabs, from the time of Mohammed down to a period contemporaneous with that when Dante flourished, were conversant with astronomy in no small degree, and that, from the wandering habits of the Arabs, it is very probable that their personal knowledge of the principal stars extended considerably south of the horizon of the Mediterranean, or of Arabia. The "four stars ne'er seen before" may, or may not, be identical with the four chief stars in the Southern Cross. If not, possibly they may refer to Canopus, Achernar, Alpha and Beta Centauri, all stars of the first magnitude; or perhaps, as we have stated above, the idea has resulted from a poetical fancy with which, in modern times, has been associated the Southern Cross. Many are inclined to believe that the last interpretation is the true one; but there are also many who think that a traditional knowledge of the existence of other large stars, unseen by the inhabitants of Europe, had been imported into northern countries by the wandering Arabs of Africa.

ERIDANUS.

Although Fluvius Eridanus, or the River Eridanus, is an immense constellation, extending from Orion in the east, to Cetus in the west, and to a point considerably below the horizon of London, it contains but very few prominent stars visible in northern latitudes. Its principal star, Achernar, is, however, a brilliant member of the first-class stars visible in the southern hemisphere. Beta Eridani, the second star, is of the third magnitude, and is near the south-western confines of Orion, north-west of Rigel, in the direction of Aldebaran.

Eridanus is one of the ancient constellations, thirty-four stars being included in the catalogue of Ptolemy. When this river was placed among the stars, it was intended by the Egyptians to typify the sacred Nile, but the Greeks gave it the name of Eridanus. Ptolemy, however, terms it merely as the asterism of the river. This constellation has occasionally been represented as a reclining female, and in the ancient manuscript of Cicero's *Aratus*, it is made to assume the form of a river-god, with his urn and other aquatic apparatus. In all celestial maps and globes of a comparatively modern date,

Eridanus is represented as a winding river, commencing in the western foot of Orion, and terminating near the constellation Phœnix, the brilliant star Achernar marking the southern boundary of the river.

HYDRA.

Hydra, sometimes called Serpens Aquaticus, the Water Snake, is one of the longest constellations either in the northern or southern hemispheres. It extends more than one hundred degrees from west to east, the head of the reptile being south of Cancer and west of Sextans. Its body takes a winding course eastward below Leo and Virgo, as far as Libra. Hydra proper includes the small asterisms of Crater and Corvus, a subdivision which is found convenient. The different sections are generally known, one as Hydra, then Hydra et Crater, Hydra et Corvus, and again Hydra. Admiral Smyth observes that " mythology calls it the Lernæan serpent; later astrologers, taking the name literally, see in Hydra the flood, in Corvus Noah's raven, and in Crater the cup out of which the patriarch sinned with the juice of the grape." The principal star in this straggling constellation is Alphard, or Alpha Hydræ, of the second magnitude, placed in a district where there is a total absence of large stars. This dearth of objects, greater than the fourth magnitude, gives Alphard much more prominence in the heavens than it is fairly entitled to. It is, however, situated in the heart of the serpent, and is known sometimes as Cor Hydræ. Its position in the heavens is very easily detected, first by its isolated appearance; secondly by drawing a line from the next two stars to the Pointers in Ursa Major, southwards through Gamma Leonis and Regulus in the Sickle; and thirdly by the aid of the astronomical rhymester.

> " Thro' Cancer's sign, whence no bright stars
>   Distinguish'd light impart,
> Pollux from Castor leads you down
>   To hideous Hydra's heart."

LEPUS.

Lepus, the Hare, joins Orion on the south, and although it is a small constellation, yet it is one of the original forty-eight asterisms of the ancients. It is said to have been placed in the heavens immediately below Orion as an emblem of caution and quickness of movement. Its principal star, Alpha Leporis, is in the body of the animal, the position of its ears being indicated by four smaller stars just below Rigel.

> " Orion's image, on the south has four stars—small but fair ;
> Their figure quadrilateral points out the timid Hare."

LUPUS.

Lupus, the Wolf, contains several stars of the third magnitude, but they are so intermixed with others belonging to Centaurus, that it is not easy to separate the two constellations. Lupus is one of the asterisms known to the ancients, and some of its stars are visible in the latitude of London when near the meridian. It has Centaurus on the south and west, Scorpio on the north, and Scorpio and Ara on the east. Lupus is generally represented in advance of Centaurus, who is pricking the Wolf with a spear.

ORION.

Orion is perhaps the finest agglomeration of stars to be found in any portion of the heavens. It is also one of the best known of the constellations. Its form is something of a quadrilateral, in the centre of which are three stars of the second magnitude, known as Orion's belt. These three stars have been also designated as the gold grains or spangles of the belt; but in former times they received the names of Jacob's Staff, the Golden Yard of Seamen, the Three Kings of Soothsayers, besides several others. They point on the one side to the bright star Sirius, and on the other to the red star Aldebaran, and the Pleiades. Betelgeuse and Rigel, two of the stars in the quadrilateral, are of the first magnitude. In this constellation, about one hundred stars are visible to the naked eye, none of which are smaller than the sixth and a half magnitude. Excepting the two most brilliant, there are four stars of the second, and five less than the second, but greater than the fourth magnitude. Orion can be seen all over the world, and is a favourite constellation in all countries. Its figure, belt, and pendant, as marked out by the stars, cause it to be easily recognised.

Orion has been mentioned by name by several of the old Greek and Roman writers. Modern hero-worship has been carried to such an extent that it has been suggested to change this name for that of noted individuals. For instance, in our own country it has been proposed to give the constellation the name of Nelson; while in 1807 the University of Leipsic actually resolved that all the stars forming the belt and sword of Orion should henceforth be known only by the name of Napoleon. The old appellation is, however, too much engrafted into the minds of all to permit a change of this kind, and the name of Orion will most probably be retained for many ages to come. Admiral Smyth, so well known as an amateur astronomer and antiquarian, has remarked that "both the Septuagint and the Vulgate call it Orion, according to the Greeks and Romans. It is mentioned in Job, Ezekiel, and Amos; and some persist that it represented Nimrod, as mighty a hunter as Orion, and the author of the post-diluvian heresy. From his terrible and threatening gesture, as much as from his time of rising, he was held to portend tempests and misfortune, and was therefore much dreaded by the mariners of yore."

A small quadruple star, Theta Orionis, visible as one object below the belt of Orion, is the centre of one of the finest nebulæ in the heavens. When viewed through a telescope, this wonderful nebula has been likened to a fish's head, to which it certainly bears a resemblance. With a twenty-foot reflecting telescope at Slough, Sir John Herschel could not compare it to anything better than a curdling liquid, or a surface strewed over with flocks of wool, or to the breaking up of a mackerel sky. It also appeared to him that the mottling of the disk of the Sun is something similar to this great nebula, although the granular look was decidedly coarser, and the intervals darker. The woolly flocks, instead of being round, were drawn into little wisps. No trace, however, could be perceived of its being composed of stars, the aspect being altogether different from that of the nebulæ which have been resolved into stars. The Earl of Rosse, with a telescope still more powerful, has seen little more than that described by Sir John Herschel. The light of this nebula has been examined with the spectroscope by Mr. Huggins, who found that, after passing through the prisms, it remained concentrated into a spectrum of three bright lines. Lieut. John Herschel has since detected a fourth line in the more refrangible part of the

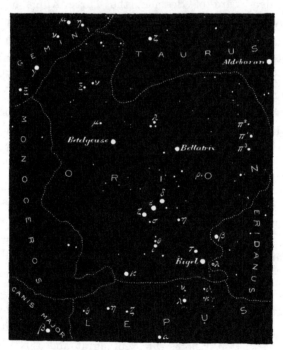

MAP OF THE CONSTELLATION ORION.

spectrum, but it is extremely faint. He also saw slight indications of a continuous spectrum. The existence of a fourth line has been confirmed by Lord Rosse and others. A spectrum of this kind is known to be exhibited only when matter in a gaseous state is rendered luminous by heat. From the positions of the four bright lines in the spectrum, it has been inferred with great probability that hydrogen and nitrogen, with a third substance not yet recognised, are present in the great nebula in Orion and in other nebulæ exhibiting a similar spectrum.

Brilliant as the constellation Taurus is with respect to the number of visible stars, it cannot in any way equal Orion for the magnitude of its components. In the latter, the universally known three stars in the belt shine conspicuously by reason of their close relationship, as well as by their equal lustre. If to these we add the four forming the quadrilateral of Orion, we have the most attractive stellar group to be found in the sky visible in Europe. Betelgeuse and Rigel are two of our brightest stars. Bellatrix and the three glittering gems in the belt are a magnitude smaller.

In the preceding diagram, we have given a representation of Orion, including the names of all its principal members, in order that the reader may be able to make himself acquainted with the individual stars of this favourite group as they appear in the sky. It will be perceived, by an inspection of the diagram, that the great brilliancy of the group consists in the quadrilateral and its inclosures, which unitedly form so great a contrast to the surrounding space, in which the paucity of stars is very marked, especially to the east, or left-hand side of the diagram occupied by Monoceros. The positions of the principal stars in the symbolical figure of Orion are as follows:—Betelgeuse on the right shoulder, Bellatrix on the left shoulder, Rigel on the left ankle, Kappa on the right knee, and the three stars Delta, Epsilon, and Zeta on the girdle or belt around the waist of the giant warrior.

In ancient times the positions of the principal stars only were recorded. Ptolemy, however, includes thirty-eight in his catalogue; Flamsteed increased the number to seventy-eight; and Bode, by inserting a smaller class of objects, has given the positions of three hundred and four. This last number expresses but very feebly what can be seen even in small telescopes, for it has been asserted that the Capuchin de Rheita, with a binocular instrument, counted more than two thousand minute objects within the boundaries of Orion.

PISCIS AUSTRALIS.

Piscis Australis, the Southern Fish, is a small constellation south of Aquarius. It is one of the ancient forty-eight asterisms, and is distinguished principally by the brilliant star Fomalhaut in the mouth of the fish. In the autumn evenings Fomalhaut is a marked object even in the latitude of London, although there it is only seen at eight degrees above the horizon. In more southern latitudes it is one of the most prominent objects. At the Cape of Good Hope, Australia, and in similar latitudes, it passes the meridian within a few degrees of the zenith.

TOUCANA.

Toucana, the American Goose, is a small circumpolar constellation situated nearly in opposition to Crux Australis, in relation to the south pole, the intervening space between these two asterisms being remarkable for the almost perfect absence of any star of reasonable magnitude. When between the zenith of our southern colonies and the pole, Toucana has Phœnix and Grus on the north, Pavo and Indus on the west, Hydrus on the south, and Eridanus on the east. The brilliant star Achernar, which marks the southern extremity of Eridanus, is near the tail of the bird; and almost due west of it, all the principal stars in Toucana may be detected in the August diagram of the sky of the southern hemisphere, looking south. This constellation is on the Antarctic circle, and contains about nine stars visible to the unassisted eye. The largest, Alpha Toucanæ, is about the third magnitude. Being so near the south pole, Toucana, in common with all the constellations in this portion of the southern heavens, was unknown to the ancients.

TRIANGULUM AUSTRALIS.

Triangulum Australis, the Southern Triangle, is of small extent, but its three principal stars, to which the asterism owes its name, are very prominent, and are frequently called "the triangle stars." Although this constellation consists of very few objects, yet its principal star is of the second magnitude, while the other two are of the third. They can be easily identified between Pavo and the two bright stars in the fore-feet of the Centaur. Their positions are always given in the index-maps. Between the Triangle and Alpha and Beta Centauri, Circinus, or the Compasses, has been inserted by the modern constellation-makers, and another small asterism, Apus Indica, or the Bird of Paradise, has been formed out of the few stars between Triangulum and the pole. Like Toucana, this constellation is situated on the Antarctic circle.

The remaining constellations south of the zodiac have not any special interest attaching to them, although several of them contain stars of the second and third magnitudes. Three of these stars are prominent isolated objects, and are very easily detected in the heavens. They are respectively the principal stars in Grus, or the Crane, Pavo, or the Peacock, and Phœnix. Dorado, or the Sword Fish, Ara, or the Altar, Indus, or the Indian, and a few others, have several moderately bright stars, but none of them exceed the third magnitude.

GENERAL NOTES

ON

THE FIXED STARS.

"How distant some of the nocturnal suns!
So distant, says the sage, 'twere not absurd
To doubt, if beams set out at Nature's birth
Are yet arrived at this so foreign world;
Though nothing half so rapid as their flight."

YOUNG.

GENERAL NOTES ON THE FIXED STARS.

THE VIA LACTEA, OR MILKY WAY.

THE course of the Via Lactea, or Milky Way, is generally laid down with great precision in all celestial maps, but the reader will also be able to gather some idea of its position, in relation to the stars, by the series of diagrams illustrating former chapters of this work. This very remarkable nebulosity extends over a vast portion of the celestial sphere, diverging, at a certain point, into two branches, which afterwards re-unite. To the eye it has the general appearance of a diffused milky light, but of variable intensity. When viewed, however, with a very powerful telescope, it is seen to consist of innumerable stars, so crowded together, at such immense distances from us, that their combined light only produces to the naked eye that nebulous appearance by which it is distinguished. The Milky Way is inclined to the celestial equator about sixty-three degrees, which it intersects in the constellations Monoceros and Aquila. Its breadth is very irregular, in some parts being only three or four degrees wide, while in others it spreads over from twelve to sixteen degrees. Starting northwards from the constellation Scorpio, it traverses in succession Aquila, Vulpecula, Cygnus, Cassiopeia, Perseus, Auriga, between the feet of Gemini and the horns of Taurus, and then over the club of Orion to Monoceros and Canis Major. From Canis Major, it enters Argo Navis nearly on the southern tropic, soon after which it descends below the horizon of London. Here it subdivides, but the principal stream diffuses itself broadly, and then assumes a fan-shape appearance. A wide gap takes place at this point, but the Milky Way recommences with a similar fan-like assemblage of branches, which all meet at no great distance from the great nebula in Argo Navis. After crossing the hind feet of Centaurus, it enters Crux Australis by a narrow kind of isthmus, when it soon expands, inclosing most of that constellation. The nebulosity is here exceedingly intense. Leaving this attractive portion of the southern sky, it passes onwards first over the two chief stars in Centaurus, then through Ara, Scorpio, Sagittarius, into Aquila, where one of the most conspicuous parts of the Milky Way visible in Great Britain is situated. The two Magellanic clouds, the positions of which are indicated in the diagrams, are apparently composed of the same materials as those forming the Milky Way. They appear as ill-defined patches of irresolvable nebulæ, of nebulæ capable of being partially resolved, and of others which can be easily separated into perfect stars, like

187

portions of the Milky Way. They are of an oval form, situated in a locality by no means rich in stars above the sixth magnitude. These nebulous clouds are distinctly visible on clear moonless nights, but the position of the Nubecula Major can be observed with the unassisted eye even at full Moon. From the earliest ages the Milky Way has maintained the same relative position in the heavens with respect to the stars.

The multitude of minute objects seen in the Milky Way through such instruments as Sir William Herschel's forty-foot reflecting telescope, or with the still greater reflector of the Earl of Rosse, is one of the most marvellous exhibitions of stellar glory with which we are acquainted. On such occasions the stars are scattered over the field of view like glittering dust on the dark ground of the sky. From this we can clearly understand that the poet Milton was not writing pure imaginative thoughts when he explained this celestial girdle as—

> "A broad and ample road, whose dust is gold
> And pavement stars, as stars to thee appear,
> Seen in the galaxy, that milky way,
> Which nightly, as a circling zone, thou seest
> Powder'd with stars."

The variability of the light of the Milky Way can be perceived in a moment on any clear moonless night, when favourably situated in the visible sky. The most brilliant part of the northern half is that which passes through Aquila and Cygnus; but this is exceeded in the southern hemisphere, where the magnificence of the Via Lactea is considerably heightened by the proximity of a large number of very conspicuous stars, including those in Scorpio, Centaurus, Crux Australis, or the Southern Cross, and others. Humboldt noticed that if, in some large portions of the Milky Way, the light is uniformly distributed, there come immediately afterwards other regions where the most brilliant parts alternate with others comparatively free from stars, giving the appearance of an irregular luminous celestial network. In certain portions of this remarkable nebulosity, perfectly obscure places are found in which it is impossible to discover a single object, even down to the eighteenth or twentieth magnitude. For example, Sir John Herschel has remarked, that in the midst of a brilliant part of the Milky Way, near the Southern Cross, "surrounded by it on all sides, and occupying about half its breadth, occurs a singular dark, pear-shaped vacancy, so conspicuous and remarkable as to attract the notice of the most superficial gazer, and to have acquired among the early southern navigators the uncouth but expressive appellation of the *coal-sack*. In this vacancy, which is about eight degrees in length, and five degrees broad, only one very small star visible to the naked eye occurs, though it is far from devoid of telescopic stars, so that its striking blackness is simply due to the effect of contrast with the brilliant ground with which it is on all sides surrounded."

When we consider that the greatest number of stars visible at any one time above the horizon is no more than two thousand, including every star as far as the sixth magnitude, it seems almost marvellous that if we were to count the same number of stars in the Milky Way we should discover that they would be contained in a very small square space of this luminous stratum. For instance, Sir William Herschel found that, in a part of the galaxy where the stars were most thinly scattered, eighty objects were, on an average, included at once in the field of view of his great telescope. Without moving his instrument, and simply allowing the stars to pass across the field by the diurnal rotation of the Earth, he found that in the course of an hour 4,800 minute stars had passed before his eye. But when the telescope was presented to a rich portion of the Milky Way, he found no less than 588 stars, and during fifteen consecutive minutes no apparent diminution in their numbers could be perceived, one constant stream of objects entering and leaving the field of the telescope in the interval. Sir William Herschel estimated that at least 116,000 stars must have passed in review before him in that short space of time. Such immense numbers are therefore contained in a narrow zone of this wonderful assemblage of stars. Most of the interesting objects termed stellar clusters are situated in or near the Via Lactea; for Herschel found that 225 are within its boundaries, while only thirty-eight had been observed in other parts of the heavens. As the Milky Way only occupies about one-twelfth part of the celestial vault, and one-ninth of that visible in this country, it has been computed that stellar clusters are fifty-four times more abundant in the Via Lactea than in other portions of the sky.

Sir William Herschel found that this stellar stratum was almost fathomless, even with his great forty-foot reflecting telescope. More recent powerful instruments have, however, revealed multitudes of stars which appeared to him only as nebulous objects; and it is very probable that future improvements in the construction of astronomical telescopes will enable the observer to penetrate still farther into these realms of space. Estimating the thickness of the Milky Way by its apparent breadth, Sir William Herschel deduced that it is about eighty times the distance of stars of the first magnitude. This stellar mass must therefore pass beyond the limit of ordinary telescopic vision. From this we may infer that not only our Sun, but every star visible to the unassisted eye, forms an integral part of the Via Lactea. The eminent Russian astronomer, the late F. G. W. Struve, has remarked that "if we consider all the fixed stars which surround the Sun as forming one great system, that of the Milky Way, we are in perfect ignorance of its extent, and we have not the least idea of the exterior form of this immense system of worlds."

MAGNITUDES OF THE STARS.

When we direct our attention to the heavens on a clear starlight winter's night, the first impression on our mind is that an almost infinite number of stars is presented to our view. This is, however, merely an optical illusion produced principally by their twinkling, and by their irregular position in the sky, for the whole extent of the heavens is too vast to be included at one time in the field of vision. Hence arises the erroneous impression that the number of stars is so great. Now, on the contrary, we know, as stated on the preceding page, that seldom more than two thousand can be perceived by an ordinary eye at once, including all as low as the sixth magnitude. Observers, however, with superior eyesight, can occasionally detect smaller objects even down to the seventh magnitude, but this exceptional vision is very unusual. There are twenty-four stars of the first magnitude in all parts of the heavens, several of them being visible only in the southern hemisphere. Those of the second magnitude number about fifty in the two hemispheres, and of the third about two hundred. Including all stars down to the sixth magnitude, or within the limits of ordinary vision, about five thousand altogether can be seen in the latitude of London during the year. But if we view the sky with a very powerful telescope, these five thousand represent but a very small proportion of the whole number of the stars; for the minute observable objects composing the groundwork of the heavens have been counted by tens of thousands, or even by hundreds of thousands.

The observed diminution in the magnitude of objects, as well as the increasing numbers contained within the field of view, as their distances increase, may be briefly illustrated as follows. Let us imagine a person standing in the middle of a forest, surrounded by trees in every direction. Those nearest to him would be few in number and the trunks comparatively large; but if he were to take the next circuit of trees outside those around him, the visible trunks would be increased in number, but their dimensions would appear smaller. Proceeding onwards in this manner, the trunks of the trees would at last be very numerous indeed, but their apparent size would bear no comparison in magnitude with those near the observer. But still these apparently small distant trees might be really considerably larger than any in the whole forest. We will now substitute the stars for the trees. For the sake of analogy, let us now suppose that the observer on the surface of the Earth is situated in the centre of a forest of stars, of indefinite extent; those few which are nearest to our own system would appear large and bright, and we distinguish them as of the first magnitude; those which are farther removed from us would appear in greater numbers, but with less intrinsic brightness— these we call of the second magnitude; those which are still farther from us would be considerably increased in number, but their magnitude would appear much smaller. If we continue increasing the distance, the decrease of brightness will be in inverse proportion to the increased number of the stars, till we are stopped by the limit of vision. With telescopic aid, the observable stars are too numerous for any accurate determination of their numbers. M. Argelander, a zealous German astronomer, has, however, actually published a catalogue of the exact positions of no fewer than a quarter of a million of stars greater than the tenth magnitude. These numbers nevertheless fail to represent properly the boundless extent of the stellar universe; for every improvement in the construction of astronomical telescopes unfolds to the view of the astronomer hundreds of thousands which had never been seen before. Sir John Herschel remarks that "beyond the limits of unaided vision, telescopes continue the range of visibility; and magnitudes, from the eighth down to the sixteenth, are familiar to those who are in the practice of using powerful instruments; nor does there seem the least reason to assign a limit to this progression; every increase in the dimensions and power of instruments, which successive improvements in optical science have attained, having brought into view multitudes innumerable of objects invisible before; so that, for anything experience has hitherto taught us, the number of the stars may be really infinite, in the only sense in which we can assign a meaning to the word."

SCINTILLATION OF THE STARS.

Perhaps one of our first practical lessons in sidereal astronomy consists in noticing that peculiar apparent motion in stellar objects, so evident on a brilliant starlight night in winter, known as the twinkling or scintillation of the stars. This phenomenon, with which most of our readers have been acquainted from early youth, by the nursery rhyme, "Twinkle, twinkle, little star," has occupied the attention of scientific men for a long period; among others, Aristotle, Ptolemy, Tycho Brahé, Galileo, Kepler, Hooke, Newton, Young, and Arago. To the unassisted eye, it consists of very rapid changes in the intensity of the lustre of the stars. These changes are also frequently accompanied by corresponding variations in colour, observations of which have been recorded by more than one astronomer. Forster, in 1824, not only noticed the variability of colour, but he endeavoured to obtain an idea of the law by which the changes took place.

One of the popular notions by which we distinguish a planet, consists in the comparative absence of any scintillation of its light, which consequently shines with a much more steady lustre than that of the fixed stars. But twinkling is not always a sure distinction between the light of the fixed stars and planets; for, in certain conditions of the atmosphere, the latter have been known to scintillate more or less, and the phenomenon is also much more observable in the fixed stars on some nights than on others. Stars of the first magnitude twinkle much more than those of the second, while in the smallest stars visible to the unassisted eye, the scintillation is altogether inappreciable. Many writers have given explanations of the cause of twinkling, each differing in many respects from the others, and even at the present time considerable differences of opinion exist. Some have accounted for the phenomenon by the undulatory theory of light, by which the direct rays from the star reach the eye at regular and successive intervals of time, causing the object alternately to appear and disappear. M. Arago considered that the scintillation of the stars is nothing more than a rapid change in their intensity and colour originating in our atmosphere, in which the progress of the stellar rays is interfered with by the unequal heating, density, or humidity of the different strata. In his opinion the principal cause of the scintillation may be supposed to arise, therefore, from the unequal refraction, or bending, of the rays of light as they pass through aërial currents of different temperatures and densities. That this is probably the case, is evident from the variability of stellar twinkling depending on the distance of the stars from the horizon. For example, scintillation is generally much more visible in stars at a low altitude, where the density of the atmosphere is always the greatest, than in or near the zenith, where the least density prevails. This law of twinkling, according to the altitude of the object, is not, however, universal, for several of the principal fixed stars, on account of the nature and peculiarity of their own light, vary considerably in the intensity of their scintillations independently of their position in the heavens. Procyon and Arcturus are known to twinkle much less than Vega, the brilliant bluish-white star in Lyra. Kaemtz states that " planets scintillate less than stars, because as the latter appear to us as points, the least displacement, were it only a few seconds, would be sensible to our eye. The planets having a visible disk, it is more difficult to appreciate their apparent change in volume ; however, through telescopes we frequently see the edges scintillate, especially if they are near the horizon." Aristotle curiously explained the phenomenon as the result of a mere strain of the eye, for he says " the fixed stars sparkle, but not the planets ; for the latter are so near, that the eye is able to reach them ; but in looking at the fixed stars the eye acquires a tremulous motion owing to the distance and the effort."

M. Wolf, Astronomer at the Imperial Observatory of Paris, made in 1867 some observations of the spectra of the stars at a time when the scintillation appeared very great. He has noticed on frequent occasions several series of broad bands or waves pass from one end of the continuous spectrum to the other, which apparently confirms the changing colour of the stars, according to the interference theory of M. Arago. On the other hand, M. Respighi, an Italian astronomer, who has also observed the wavy motion from one end of the stellar spectrum to the other, differs in many respects from the conclusions of M. Wolf, and throws some doubt on the accuracy of the theory of Arago. M. Respighi considers that the cause of scintillation arises from real and momentary deviations in the positions of different coloured rays produced by the changing state of the atmosphere, and that when these rays are abstracted from the images of the stars, they would naturally exhibit continual variations in their colour and intensity. He has also noticed that the motion of the undulations is from the red end of the spectrum towards the violet end for the stars in the west, and on the contrary, from violet to red for the stars in the east, while on the meridian the motion is more or less irregular. The undulations are seen with much more distinctness in stars near the horizon than in those at a great elevation.

M. Respighi, from a due consideration of the phenomena observed by him, concludes that the regularity of the motion of the undulations in the spectrum, taking into account the opposite movements for the stars near the eastern and western horizons, shows clearly that the waves or atmospheric strata are brought successively on to the luminous rays transmitted by the stars, not so much by any accidental motion of the atmosphere, as by its general movement, either ascending towards the west or descending towards the east, resulting from the diurnal motion of the Earth. " When the luminous rays," he observes, " traverse the lower regions of the atmosphere, where, in consequence of the differences of temperature and of the unequal condensation of the aqueous vapour, the atmosphere itself is found in a state of sensible heterogeneity, the condensed or rarefied atmospheric strata would be brought, by the diurnal rotation of the Earth, for the stars in the west first on the red rays, and afterwards on the more refrangible rays as far as the violet ; and, on the contrary, for the stars in the east, at first on the violet rays, and finally on the red. The effect of this is that the corresponding undulations on the spectrum would travel from the red to the violet for the west stars, and from the violet to the red for the east stars, precisely in the same manner as shown by the observations."

M. Dufour, who made an extensive series of observations on stellar twinkling at Morges, Switzerland, found that the phenomenon varies frequently in the same locality, from one day to another. But it increases or diminishes proportionally for all the stars, excepting those near the horizon, where the twinkling is always large. It has also been observed to increase during the time of twilight, and when clouds are in the sky driven rapidly before the wind. During those nights in which the scintillation was very marked, M. Dufour noticed that the stars in all directions, including the zenith, were affected; but on nights when the phenomenon was less decided, all the zenithal objects shone steadily. In tropical countries, scintillation is but seldom observed in stars at a high elevation above the horizon, and then only to a very limited extent. Humboldt remarks that in Peru stars scintillate when near the horizon, but not at more than twenty degrees above it. Garcin, in a letter to M. Reaumur, published in the *Histoire de l'Académie des Sciences*, 1743, states that "in Arabia, in spring, summer, and autumn, the inhabitants sleep on the roofs of their houses. It is impossible to describe the pleasure experienced in contemplating the beauty of the sky, the brightness of the stars, and their apparent motion from east to west, while thus lying in the open air. The light of the stars is pure, steady, and brilliant; and it is only in the middle of winter that a slight degree of scintillation is observed."

Observations on the scintillation of the stars have been made by several observers in elevated positions on the Earth's surface, especially by Saussure in the mountainous districts of Switzerland, and Professor C. Piazzi Smyth on the Peak of Teneriffe. At the latter place, Professor Smyth was much struck by the quiet and steady planetary light of the stars, and was inclined at first to believe that there was no scintillation; but he soon found that this phenomenon even existed at his elevated station of 10,702 feet, although to a much smaller extent than at the foot of the mountain. Dr. Tyndall noticed in 1859, from the Grands Mulets, on Mont Blanc, at an altitude of nearly twelve thousand feet, that when Capella first appeared near the horizon on the evening of August 13th, the star scintillated very distinctly, but that at 2 A.M. on the morning of the 14th, the twinkling was scarcely perceptible. From this observation of Dr. Tyndall, we may conclude that, when viewed from the summit of Mont Blanc, the stars shine with a steady light. During a residence at the hospice of St. Bernard in the summer of 1856, M. Dufour also found that the scintillation was very trifling. Whether this absence of the phenomenon at such great elevations occurs at other seasons of the year, we have no recorded observations to show.

The occasional twinkling of the planets consists only of a slight tremulous motion of their disks when near the horizon. It has been noticed principally in Mercury, Venus, and Mars. But in such cases, the phenomenon is so difficult to observe, that practically it may be concluded that, to the naked eye, any displacement resulting from it is too minute to be perceived by any but practised observers.

s

COLOUR OF THE STARS.

No one can view the heavens with the naked eye for any length of time, without noticing that the colour of the principal stars is not the same, but that many are white, or bluish white, some yellowish, and others red. These different tints arise most probably from the materials of which the envelopes, or photospheres, of the various stars are composed. If, however, we wish to examine this subject minutely, the telescopic observation of double stars will enable us to view, in the highest perfection, the brilliant contrasting colours exhibited by several of these interesting and popular objects. With the unassisted eye, the variation in colour of the brightest stars is very distinctly marked, especially in the white stars Sirius, Alpherat, Vega, Deneb, and Regulus; the yellow, or pale orange stars, Rigel, Procyon, Polaris, Kocab, and Altair; the orange-red Betelgeuse and Pollux; and the ruddy stars Aldebaran, Antares, Arcturus, and Fomalhaut. The only colours mentioned by the ancients, whose experience in stellar observation was solely derived from unaided vision, were white and red. Some of these anciently-recorded colours of the stars do not, however, agree with the colours of the same stars at the present day, which leads us to suppose that some physical change has taken place in the constitution of their external envelopes. We have previously alluded to the probable change of the colour of Sirius from red to white since the days of Ptolemy, who also included Pollux and Betelgeuse among his list of fiery-red stars. The change of colour has, however, not been so decided in Pollux and Betelgeuse as in Sirius; for Pollux and Betelgeuse have still a rosy tint, although it is too faint to take them out of the class of orange-coloured objects. Mariotte, in 1686, in his treatise on *Colours*, was the first person who made any mention of blue stars. He considered that the origin of blue stars was owing to "their freedom from exhalations as well as from their less intrinsic brightness." True blue single stars are not, however, common, but bluish white are plentiful enough. Mr. Dunlop, observer at the late Sir Thomas Brisbane's Observatory, Paramatta, New South Wales, noticed a stellar mass, in which every member was blue, and also a bluish nebulosity. Nothing of a similar kind has been seen in the northern hemisphere. But several of the components of the double stars are blue, and in a few cases both have a bluish tinge. Small stars of different colours are occasionally massed together in multiple stars, as in that beautiful stellar group near Kappa Crucis (the Southern Cross), in which are congregated more than a hundred stars of various colours, red, green, blue, and bluish green, giving the appearance, when viewed through a powerful telescope, of a superb collection of fancy jewellery.

But notwithstanding the brilliancy of colour in these magnificently variegated minute objects, scattered here and there in both hemispheres, the principal observations on the colour of stars have been made on the various tints exhibited by double stars, in which the colours are generally complementary. Usually, the larger star is of a yellow, or orange colour, and the smaller one green, or bluish white; but among these objects there is no shade of colour contained in the solar spectrum, which is not also represented in some one of these double stars. Admiral Smyth, in his *Sidereal Chromatics*, has given a list of 109 double stars, with the colours of each pair, as observed by himself at two epochs, separated by several years, and also by M. Sestini, an Italian astronomer. The estimations in different years generally agree with each other, but in a few cases an actual change of colour has been suspected, especially in 95 Herculis, in which both components are of the fifth magnitude. The change in the colours of this double star is so very curious, that it may be interesting to give the details of the observed variations in the colours of the two components, A and B, as recorded by Professor C. P. Smyth, in a communication inserted in *Sidereal Chromatics*. The observations were made by different observers between the years 1828 and 1862. In 1828 the component A was yellow; from that it passed to greyish, then successively to yellowish, with a blue tinge, greenish, light green, light apple green, "astonishing yellow green," and, finally, to yellow again. B in the same time passed from yellow to greyish, then successively to yellowish with a reddish tinge, reddish, cherry red, "egregious red," and, finally, to yellow again. It was the opinion of Admiral Smyth, whose experience was very great in this class of amateur astronomical work, that the variable appearance of this star is a decided instance of sidereal colour-changing. It is proper, however, to remark that the Astronomer Royal has suggested, with good reason, that the simultaneous change in the colours of the two stars is suspicious, and he considers that these apparent changes might have arisen by using different telescopes. Other stars viewed with the same telescopes, as 95 Herculis, however, exhibited no sensible change.

From Admiral Smyth's latest observations we give two or three examples of the complementary colours. In Eta Cassiopeiæ the large star is a dull white, and the smaller one lilac; in Gamma Andromedæ, a deep yellow and sea-green; in Iota Cancri, a dusky orange and a sapphire blue; in Delta Corvi, a bright yellow and purple; and in Albiero or Beta Cygni, yellow and blue. In most of the remaining stars of the list the contrasting colours are equally marked, and also in many others which are not included in it. What the effect of this variety of colour would be to the inhabitants of a satellite belonging to such a cosmical system, we have the opinion of Sir John Herschel. " It is by no means intended to say that in these cases one of the colours is a mere effect of contrast, and it may be easier suggested in words than conceived in imagination, what variety of illumination *two suns*—a red and a green, or a yellow and a blue one— must afford a planet circulating about either; and what charming contrasts and 'grateful vicissitudes'—a red and a green day, for instance, alternating with a white one and with darkness—might arise from the presence or absence of one or other, or both, above the horizon."

PROPER MOTIONS OF THE STARS.

We have frequently alluded to the term "fixed," as an epithet given to distinguish the great majority of stars from the Sun, Moon, and planets, whose apparent positions in the heavens are continually changing sensibly from day to day, and, in the case of the Moon, from hour to hour. The so-called fixed stars, so far as can be measured by the unassisted eye, however, never alter their relative positions with respect to each other, and appear as if they were attached to the celestial sphere. They are observed to rise in the east and set in the west, from day to day, and from year to year, without any sensible change in their general aspect, excepting only that due to the seasonal variation produced by the apparent motion of the Sun on the ecliptic, as briefly explained in the description of the "Midnight Sky at London" for March. This apparent absolute fixity of the stars in space was supposed by the ancients to be real, and it was not till the year 1717 that Dr. Halley, owing to the greater accuracy of modern astronomical observations, noticed that the positions of the three bright stars, Sirius, Aldebaran, and Arcturus, were from one-third to half a degree more southerly than those recorded by Ptolemy on the authority of some observations of these stars made by Hipparchus, 130 B.C. At first it was naturally supposed that the discrepancy arose from errors in the observations of the ancient astro-nomer, but Dr. Halley considered that the observed differences between the ancient and modern positions of these three stars were more likely caused by a peculiar motion of the stars themselves. This explanation or opinion of Dr. Halley has been completely confirmed by the more precise observations of the present century. In truth, it has been found for a certainty, that a very large number of stars have their own "proper motion," some of greater extent than others. The right ascension and declination of upwards of 3,000 stars, observed by Dr. Bradley between 1750 and 1762, have been compared with modern observations of the same stars, and the annual proper motion of each accurately determined. In a few stars these peculiar displacements reach to a very sensible amount, the largest being nearly eight seconds of arc annually. In the binary star, 61 Cygni, this proper motion consists of rather more than five seconds, and it is found that the two stars composing the binary system have the same amount of annual displacement. This equal annual angular motion is a convincing proof of their physical connection, independently of their revolution around each other, as we have explained more fully in a previous chapter.

The subject of the proper motions of the "fixed stars" has been in many ways extremely interesting to astronomers, as it is by the systematic discussion of these appa-rently small quantities, that the existence of a supposed motion of the solar system in space has been determined. Sir William Herschel, in 1783, was the first who drew the attention of other astronomers to the probable existence of this proper motion of the solar system. His research led him to point out Lambda Herculis as the direction in the heavens towards which the Sun was moving. One argument in favour of the problem, and a very strong one, is that successive astronomers with different data and methods of research have, without exception, found the solar motion directed towards the same point.

A thorough investigation of the subject by the Astronomer Royal has, however, thrown some doubts on the reality of this rather romantic astronomical problem, or at all events on some of the received notions respecting it. A more extensive inquiry made in 1863 by the author of these pages, by applying Mr. Airy's formulæ to the proper motions of 1,167 stars, confirmed the doubts expressed by the Astronomer Royal, although the deduced direction of solar motion agreed with that found by Sir William Herschel, and the velocity with that determined by M. Otto Struve. But although the supposed movement of the Sun in space thus appears to account for only a small part of the star displacements, yet so distinguished an astronomer as Sir John Herschel has remarked that it is not surprising that such should be the case; for in his opinion it could not be expected that any movement assigned to the Sun would account for more than a very small portion of the observed proper motions. He says:—" But what is indeed astonishing in the whole affair is, that, among all this chaotic heap of miscellaneous movement, among all this drift of cosmical atoms, of the laws of whose motions we know absolutely nothing, it should be possible to place the finger on one small portion of the sum total, to all appearance undistinguishably mixed up with the rest, and to declare with full assurance that this particular portion of the whole is due to the proper motion of our own system."

The following are the points in the heavens which the different researches have assigned as the direction of solar motion :—

	R. A.	N.P.D.
W. Herschel	260° 34′	63° 43′
Argelander	257 35	53 57
Lundahl	252 53	75 34
Mädler	261 38	50 6
Galloway	260 1	55 37
O. Struve	261 22	52 24
Airy	259 12	57 53
Dunkin	262 29	61 2

It has been asserted by Dr. Mädler that his investigation gives some evidence of the possibility of the Sun and its system revolving around a central body at a rate of about 150 millions of miles in a year. This central Sun of the universe has, according to Dr. Mädler, been supposed to be the star Alcyone, the brightest object in the Pleiades group. There is, however, very little grounds for the conclusion adopted by this celebrated Russian astronomer; for although the subject is very interesting and romantic, the proof of the existence of a central Sun for the whole universe must still be considered an unsolved problem belonging wholly to the speculative branches of astronomy. Many more centuries must elapse before sufficient observations can be obtained to enable any one to speak positively on the question. The very small quantities representing the observed proper motions of the stars, require to be completely verified by most accurate observations, made at intervals of one or two hundred years, before we shall be even certain that the solar system itself has, or has not, the peculiar motion in space now ascribed to it.

ANALYSIS OF SOLAR AND STELLAR LIGHT.

One of the most interesting branches of observing astronomy of late years has been the telescopic observation of the spectra of the light emitted by the Sun and stars, and the comparison of these spectra with those found by the spectroscopic observation of intensely-heated luminous vapours produced by the combustion of metals and gases. When a beam of solar light is made to pass through an ordinary prism, it is so refracted, or bent, as to be resolved into a number of divergent rays, which, when projected on a screen, form an oblong luminous band of different colours called the prismatic solar spectrum. The colours, commencing at the top, are violet, indigo, blue, green, yellow, orange, and red. The red is the least refrangible because the red ray is less turned out of its course by the interference of the prism than the others; in like manner, the violet is the most refrangible, because that coloured ray is the most refracted, or bent, out of its course. The cause which produces the prismatic spectrum was first explained by Sir Isaac Newton. If, instead of a screen, we view the sunbeam after it has been made to pass through a spectroscope, we shall see hundreds of narrow dark lines crossing the spectrum transversely from one end to the other. Some of the lines, visibly broader than the others, are distinguished by the first roman letters of the alphabet. These dark lines were first observed by Dr. Wollaston in 1802. They are, however, more popularly known as Fraunhofer's lines, on account of the care bestowed by that eminent German optician in the construction of a map of the solar spectrum in which the dark lines are shown in their relative positions. The lines have subsequently been laid down with far greater completeness, first by M. Kirchhoff, and afterwards by M. Angström.

By the researches of MM. Kirchhoff and Bunsen, who were the first to connect the dark lines in the solar spectrum with corresponding bright lines seen in the spectra of the vapours of terrestrial metals and gases in a state of combustion, astronomers have been enabled to prove, with almost perfect certainty, that the Sun and Earth contain many metals and gases common to both. Among them we may include iron, sodium, copper, magnesium, barium, nickel, calcium, and chromium, with indications of many more.

It was not the original intention of the two physicists, Kirchhoff and Bunsen, to explain the origin of the dark lines, their attention being chiefly directed to the determination of the position of the various bright lines found in the spectra of the different metals, using the dark lines in the solar spectrum solely as points of reference. Their first operation was to test an assertion made by Fraunhofer in 1814 that the position of two of the black lines in the solar spectrum corresponded in refrangibility to the two bright lines known to exist in the spectrum of the vapour of sodium. By mingling the spectra coming from a flame coloured with sodium, and from the light of the Sun, Kirchhoff observed the bright sodium lines superposed upon the dark lines in the solar spectrum. In the paper by Kirchhoff, published in the *Mémoires de l'Académie de Berlin* for 1861, the details of the experiments carried on by himself and M. Bunsen are given with great completeness. "I caused," he observes, "a direct solar ray to pass through the flame of sodium placed before the slit of the spectroscope, and I saw, to my great surprise, the line D appear with an extraordinary intensity. Instead of a direct solar ray, I now used the Drummond light (lime rendered incandescent by the combustion of hydrogen), whose spectrum, like all solid or liquid bodies heated to incandescence, does not exhibit the dark lines. By passing the light through a flame charged sufficiently with marine salt, dark lines were immediately substituted for the bright lines of the sodium. Employing, instead of the Drummond light, a wire of platinum rendered incandescent by means of a flame, and raised to a temperature nearly at the point of fusion by the action of an electric spark, I observed the same fact." Continuing his investigations on the bright lines of vaporised metallic spectra, still using the dark lines for comparison, M. Kirchhoff on one occasion examined critically the bright lines in the spectrum of iron. "Judge," remarks Mr. Roscoe, "of the astonishment of Kirchhoff when he observed that dark solar lines occurred in positions coincident with those of all the bright iron lines! Exactly as the sodium lines were identical with Fraunhofer's line D, so for each of the iron lines of which Kirchhoff and Ångström have since mapped no less than 460, a dark solar line was seen to correspond. Not only had each iron line its dark representative in the solar spectrum, but the breadth and degree of shade of the two sets of lines were seen to agree in the most perfect manner, the brightest iron lines corresponding to the darkest solar lines. To those who have not themselves witnessed this coincidence it is impossible to give an adequate idea by words, of the effect produced on the beholder when, looking into the spectroscope, he sees the coincidence of every one of perhaps a hundred of the iron lines with a dark representative in the sunlight; and the idea that iron is contained in the solar atmosphere flashes at once on his mind."

One of the conclusions drawn by Kirchhoff from these experiments is that each incandescent gas weakens, by absorption, rays of the same degree of refrangibility as those it emits; or in other words, that the spectrum of each incandescent gas is reversed when this gas is traversed by rays of the same refrangibility emanating from an intensely luminous source which gives of itself a continuous spectrum like that of the Sun. The dark lines in the solar spectrum most probably arise, therefore, from the metallic rays emitted from the incandescent vapour, either in or below the ordinary photosphere, being absorbed by the cooler strata in the exterior atmosphere or envelope of the Sun, or possibly in some degree by the photosphere itself. In the spectra of vaporised metals and gases, it has been universally found that one substance gives one series of bright lines, another a different series, the spectrum exhibiting, in fact, certain lines peculiar to the metal or gas employed in the experiment. Consequently, whenever bright lines in a metallic or gaseous spectrum are found to coincide with corresponding dark lines in the solar spectrum, there can be little doubt that these terrestrial elements must form a constituent part of the great central globe of our system.

From these celebrated experiments of the two German physicists, all the additional knowledge of the constitution of the heavenly bodies which we have obtained during the last few years has mainly sprung. Mr. Huggins, Dr. W. A. Miller, and Mr. Lockyer, in England; M. Janssen, M. Rayet, M. Secchi, and others, on the Continent; have each used the spectroscope on the heavenly bodies with such a success as even M. Kirchhoff himself little anticipated. Since his observations, a whole harvest of facts has been gathered together, throwing out a new flood of light on the general structure of the sidereal universe. The constitution of the photosphere of the Sun is now being understood as it never was before; the light of the Moon, planets, comets, stars, and nebulæ, has in each case been impounded, so that we may decide from what substances it has been emitted. The analysed light of the Moon and planets being the reflected light of the Sun gives naturally the solar spectrum, but on account of local causes, each planet has special dark lines exhibited in its spectrum. For example, the spectra of Venus, Mars, Jupiter, and Saturn, give some indications of the presence of a gaseous atmosphere around them, bearing some analogy to that of the Earth, including the presence of the vapour of water. Jupiter and Saturn have, however, a certain similarity on other points not shared in by the nearer planets. The spectrum of Uranus has also something peculiar to itself. The analysis of the light of the Moon gives no trace of the existence of an atmosphere, as had been previously proved by other classes of astronomical observation.

Nearly all the stellar spectra exhibit only the dark, or absorption lines, as in the case of the Sun. The distribution of the lines is, however, very different. M. Secchi has found that the stars may be roughly classified according to their colour, by dividing them into four types. First, there are the white stars, such as Sirius, Vega, Spica, and Altair. These stars exhibit more particularly the presence of hydrogen gas at a very high temperature, while indications of other substances, such as sodium and magnesium, are also clearly shown. The second division, or type, appears to have a composition very similar to that of the Sun, and includes Arcturus, Capella, Pollux, and others. These two classes contain the greater number of the principal stars. The third division, including Betelgeuse, Antares, Alpha Herculis, and others of like colour, seems to be affected by something peculiar in their physical composition, as if their photospheres contained a quantity of gas at a lower temperature than usual. The stars in this class have generally a ruddy tint, probably owing to their light having undergone some modification while passing through an absorbing atmosphere, as is the case with the planets. M. Secchi has discovered that the spectrum of the interior of the solar spots bears a strong resemblance to that of the red stars, more particularly Antares, Mira Ceti, Aldebaran, and Alpha Herculis. The spectrum of each of these stars contains a very large number of dark lines. He considers that the direct conclusion which we may draw from this coincidence is, that these stars may owe their colour to a similar cause to that which produces the spots on the Sun. These stellar spots, under this hypothesis, ought to be much more numerous than those exhibited on the solar disk, even at the times of maximum frequency. Now, a great number of the stars in the third division are variable in their lustre. This variability may therefore be explained, if we assume that the stellar spots have a periodicity in a similar manner to that which has been observed in the Sun. The period of the solar spots is tolerably well ascertained, so the Sun may be fairly called a variable star belonging either to the second or the third division or type. The fourth class is composed principally of smaller objects, with spectra not very dissimilar from the third type, only less brilliant. A few stars have been found whose spectra exhibit bright lines indicating the presence of hydrogen and other gases in a state of incandescence. Three of these are in Cygnus. Gamma Cassiopeiæ and the variable star R Geminorum also give strong evidence of the presence of luminous hydrogen gas in their photospheres. The celebrated temporary star in Corona Borealis which burst forth so suddenly in May, 1866, was clearly owing to a combustion on a large scale of escaped gas of a similar kind.

Mr. Huggins was the first observer who succeeded in obtaining satisfactory spectra of the nebulæ and clusters. He has found that these faint cloud-like objects may be divided into two classes, one giving a continuous spectrum like the Sun and stars, and the other a spectrum consisting only of a few bright lines. All the nebulæ which have been resolved into stars by the aid of great optical power, and some others, belong to the first class. The spectroscope therefore enables the observer to judge which of these faint objects is composed of distinct stars, and which of irresolvable matter, or gas in a state of incandescence. The great nebula in the sword-handle of Orion, which at one time was supposed to show symptoms of resolvability when viewed under favourable circumstances with the Earl of Rosse's great six-foot reflector, is one of those whose spectrum consists of bright lines. Four have been observed in the spectrum of this well-known nebula, and some other bright lines have been suspected by the Earl of Rosse and Professor Winlock. In nebulæ where the spectrum is proved to consist of bright lines only, it is not possible for the nebulous matter to be resolved into distinct stars, however powerful the telescope employed may be.

The nuclei of comets shine by their own light. The nature of that light places these erratic members of the solar system almost within the class of nebulæ, or rather among the third type of stars. In 1864, M. Donati, of Florence, observed a spectrum of bright lines in a small comet; and subsequently Mr. Huggins has observed similar spectra in four others, two of which, Brorsen's and Winnecke's comets, appeared in 1868. Mr. Huggins considers that his observations seem to show decidedly that the principal portion of the light from the heads of the comets is very differently constituted from solar light, and, therefore, it cannot be the Sun's light sent back to us by *ordinary* reflection from the cometary matter. The spectrum of Winnecke's comet appeared very similar to that of the vapour of carbon when viewed in direct comparison with the spectrum of olefiant gas. Mr. Huggins remarks that "the obvious and apparently well-founded conclusion from these observations would be that the cometary matter from which the light comes, consists of the luminous vapour of carbon."

This new branch of astronomical research, opened since the experiments of MM. Kirchhoff and Bunsen, has already been productive of important results, some of which have been proved with remarkable precision. Other results, although still retaining much of a speculative character, contain the elements of truth, only requiring perhaps many years' close observations before any satisfactory confirmation may be hoped for. As a subject for the investigations of the astro-physicist, the examination of the luminous spectra of the heavenly bodies has proved a remarkably fruitful one, of which only the preliminary reaping has been garnered. We are indeed as yet on the threshold of the discoveries which the use of the spectroscope, in combination with the telescope, will probably enable us to make on the physical constitution of the universe. The rich harvest already obtained points hopefully to the future, and it gives encouragement to the investigator to continue his researches on a subject in which what is well understood bears but a small proportion to that which still remains a hidden mystery even to the most gifted of human intellects. The effects of the inquiry into this new branch of astronomy will be sure to be reflected in time, not only upon highly cultivated scientific minds, but upon every admirer of those wondrous works of the Creator, whose omnipotent hand is seen in all this uniformity of construction unfolded to us by the spectroscopic analysis of solar and stellar light.

NOMENCLATURE OF THE STARS.

The division of the heavens into groups of stars, and individualising these groups by distinctive names, most probably originated at the very infancy of astronomy. That this is the case is very evident from the astronomical allusions in the Book of Job, and in other portions of the Old Testament. In the same manner as the traveller naturally attaches some name to any remarkable terrestrial object which he may have discovered, in order to distinguish it from those already known in other parts of the globe, so the first watchers of the stars would probably attempt to give names to those principal objects in the heavens which might serve as guides in their nightly wanderings over desolate tracts of country, or in the earliest voyages in which the adventurous seamen first lost sight of land. The splendid object in Canis Major would most likely be one of the first to strike the attention of the nomadic astronomer; and hence the name of Sirius was given to it as its distinctive mark. Orion, the Pleiades, Arcturus, Aldebaran, and most of the principal stars, were all evidently named about the same time. The advantage arising from the use of these old distinguishing names has been acknowledged by modern astronomers, and it would be a step in the wrong direction to make any change in their well-known nomenclature. But although it is not considered advisable to make any alterations in the names of the individual stars, yet more than one astronomer of note has expressed an opinion that stellar astronomy would be greatly advanced by a modification of the existing constellated groups. In the primitive arrangement, or parcelling out of the stars, there is no doubt that the actual aspect of the heavens was consulted, and that the configuration of the stars adopted for each constellation was supposed to bear, in the imagination of the inventors, some feeble resemblance to the form of the figure intended to be represented. For example, when the complete tail of the Scorpion is visible at a tolerable elevation, it very admirably gives a fair idea of what we consider a tail of a scorpion ought to be like. The star Lambda Scorpii, between the second and third magnitude, forms the bulb of the sting; while a smaller star, Upsilon Scorpii, may be taken for the sting itself. The group of stars in Leo called the Sickle bears a very close resemblance to that agricultural implement; while the seven popular stars in Ursa Major might be taken as the celestial representation of a plough or waggon. But of all the ideal human figures which give names to the constellations, scarcely one can be found whose outline can be detected by the stars contained within its boundaries. Sir John Herschel remarks on this subject that "Orion is the only group which can anyhow be tortured into a good symmetrical representation of a human figure: and even this, to be seen to advantage, must be viewed from a southern latitude, where the stars Betelgeuse and Bellatrix, which in our maps form the shoulders, will appear as the knees of the figure, Rigel, as the head, and Beta Eridani as one of the shoulders, forming a really superb and majestic outline, worthy of a hero or a demigod." In the original selection of the names of the constellations by the ancients, it was perfectly natural that they should be imposed from other associations than that arising from any similarity of form to particular men or animals. To a strongly imaginative people like the Greeks, gods, demigods, and heroes would probably be the first objects presented to their minds as the proper permanent denizens of the skies. This feeling did not, however, belong exclusively to any intellectual superiority of the ancient Greeks, but it is a natural consequence of an innate bias of the human mind, shared in a more or less degree by us all, and from which even the untutored savage is not altogether exempt. "The Pampas Indians," says Captain F. B. Head, in his *Rough Notes in the Pampas*, "believe in a future state, to which they conceive they will be transported as soon as they die. They expect that they will then be constantly drunk, and that they will always be hunting; and as they gallop over their plains at night, they will point with their spears to the constellations in the heavens, which they say are the figures of their ancestors, who, reeling in the firmament, are mounted upon horses swifter than the wind, and hunting ostriches."

A great deal of the confusion existing at present by the interlacing of different constellations has been caused by the introduction of so many new asterisms by Hevelius in the seventeenth century, and by subsequent astronomers. Many of these were formed in acknowledgment of court patronage. Even the celebrated Halley, who had received important favours from Charles II after the restoration, detached in 1676 a small portion of Argo Navis, containing a group of stars which were supposed to bear some resemblance to a tree, to which he gave the name of King Charles's Oak, in remembrance of the oak in which the Prince was concealed when pursued by Cromwell's soldiers after the battle of Worcester in 1651. Twenty-six small constellations were added in the eighteenth century. Nearly all these were formed of scarcely visible stars; and the study of astronomy was by no means made more easy by their introduction. These new constellations are so unsuited to the others, and their names are chosen with so little taste, that no one can consult a modern celestial globe without being confused by their numbers and insignificance. A German astronomer, Dr. Olbers, has remarked that "surely, if it were requisite that the whole heavens should be filled with constellations, they might have been chosen according to some general principle. We might have embellished the apparatus and inventions of our chemists, if, indeed, they could be embellished by them; and as the

203

ancient figures of heroes and animals must be retained, some latitude might be allowed also to astronomical instruments. But figures like the shop of the sculptor, the chemical furnace, the easel, the microscope, the air-pump, etc., have no relation to the sky, and their being mixed up with the others, is heterogeneous, disagreeable, and without any taste."

In the present advanced state of astronomical work as practised in observatories, the stars are known not always according to the constellations, but frequently according to their right ascensions and declinations, especially the smaller class of objects. Several of the principal constellations extend over several hours of right ascension, such as Serpens, Draco, and Hydra. Serpens actually passes through Ophiuchus. Seeing the many inconveniences attaching to the irregular forms of the constellations, it had been proposed by several astronomers, especially by Dr. Olbers and Sir John Herschel, that a revision of them might be made so as to divide the whole extent of the heavens into well-defined, if not equal, portions. Sir John Herschel gave much attention to the subject, and he drew up a scheme for the purpose. His great name was not, however, sufficient to enable his proposition to be accepted by other astronomers. It was generally considered that the expected advantages to be gained by a modification of the nomenclature as derived from the ancients, would not be a sufficient compensation for the loss of the old historic appellations by which the principal constellations have been distinguished from the earliest periods in the annals of astronomy. So far as regards the constellations which rise above the horizon of Europe, or at least those which were invented by the primitive astronomers, considerable difficulty would naturally occur in any revision before it could be made generally acceptable; but in the southern hemisphere, where most of the constellations are comparatively of modern origin, and have not the sanction of such venerable antiquity as the others, a revision might be more easily effected.

The ancient catalogue in the *Almagest* of Ptolemy, compiled in the beginning of the second century of the Christian era, contains very few records of stars not visible in Europe. Of the strictly southern constellations, only the following appear :—Argo Navis, Ara, Centaurus, Lupus, and Corona Australis. Achernar in Eridanus, and Canopus in Argo Navis, are the two only southern stars of the first magnitude whose names appear in that catalogue. Phœnix, although no part of it ascends above the horizon of Europe, was known to the Arabians under the appellation of the Griffin, or Eagle, from a very remote age. In the fifteenth century, when navigation was extended beyond the equator, and sailors first became acquainted with those stars in the southern hemisphere unknown to the ancients, they found it convenient and useful to adopt a similar plan of grouping the stars into constellations as in the northern hemisphere; hence the principal asterisms which immediately surround the south pole received their names. Most of the others, which are, however, of small extent, were added by Royer in 1679, by La Caille in 1752, and a few by later astronomers.

OBSERVATORIES IN THE SOUTHERN HEMISPHERE.

Soon after the invention of the astronomical telescope, the zeal and energy of more than one astronomer was directed towards the unexplored stellar regions of the south. About the same time that the influence of the members of the lately constituted Royal Society was sufficient to persuade Charles II to order the erection of a building in Greenwich Park for the observation of the Moon, for the solution of the problem of determining the longitude at sea, a young astronomer volunteered to proceed to some station south of the equator, to observe the accurate positions of the principal stars invisible in northern latitudes. Furnished with a recommendation from the king, who, with all his faults, was a great supporter of astronomical science in his day, young Halley, under the patronage of the East India Company, sailed for St. Helena in November, 1676. During a residence of two years in this island, Halley made sufficient observations to enable him to publish, in 1679, his *Catalogus Stellarum Australium*, containing, besides the positions of 350 stars, some other points of importance, especially an observation of a transit of Mercury over the Sun's disk, in the remarks on which he suggested that such observations, if made at different places on the Earth's surface, would be a good method for determining the Sun's parallax, and consequently the distance of the Sun from the Earth. In another chapter we have referred to a like determination from the transit of Venus of 1769, the true value of the observations of which has only lately been fully ascertained. It is very probable that the original idea of young Halley was never forgotten in after life by this experienced astronomer, for in 1716 he suggested that the transits of Venus over the Sun in 1761 and 1769 ought to be fully observed; and he earnestly implored astronomers not to neglect so valuable an opportunity for determining so important an element as the horizontal parallax of the Sun. Throughout his long astronomical career till his death, as Astronomer Royal, in 1742, Dr. Halley never lost an opportunity of impressing upon his younger scientific brethren the value of these observations. How his wishes were eventually carried out by his astronomical successors, especially by another Astronomer Royal, Dr. Maskelyne, need not here be further enlarged upon.

About the middle of the last century, a distinguished French astronomer, M. La Caille, proceeded to the Cape of Good Hope with the requisite instruments for making an extensive series of observations of the southern stars. During his residence in South Africa, La Caille made sufficient observations for determining the parallaxes of the Moon and of the planets Venus and Mars, and also for investigating the laws of atmospheric refraction. He carefully measured the exact length of an arc of the meridian, an operation requiring considerable skill and attention. In addition to these important works, he fixed the positions of ten thousand stars, which have been formed into a catalogue under the auspices of the British Association. All this astronomical labour was performed without the aid of an assistant, and with instruments by no means so perfect as those of the present day. The observations were made principally in 1751 and 1752, with a small telescope of 26¼ French inches focal length and one half-inch aperture, magnifying only about eight times. La Caille did not live to see the results of his labours appreciated by his fellow-observers; for he died a victim to incessant application on March 21st, 1762, while his observations were published in the succeeding year by his friend and executor, J. D. Maraldi.

Two observatories which have done good service to astronomy, have been established and dismantled in the first half of the present century. The first was situated at Paramatta, about fourteen miles distant from Sydney. When Sir Thomas M. Brisbane was appointed governor of New South Wales, he resolved to avail himself of the opportunity thus afforded for promoting the science of astronomy among the colonists in that district of the southern hemisphere. He therefore erected an observatory at Paramatta in 1822, and furnished it with a transit instrument, a mural circle, and other instruments, at the same time appointing two assistant astronomers to make a regular series of observations. A catalogue of the positions of 7,385 stars has been the result of their labours. On the departure of Sir Thomas Brisbane from the colony at the end of 1825, the observatory was transferred to the colonial government, who appointed Mr. Charles Rümker to be the astronomer in charge. In 1829 he was succeeded by Mr. Dunlop, who had made most of the observations under Sir Thomas Brisbane. The observatory is now closed, the instruments having been transferred to the Sydney Observatory or to private establishments. The second observatory which had only a temporary existence is that of St. Helena. The principal observations, and the only ones of importance, were made in 1830–32 by Mr. Johnson, afterwards Radcliffe Observer at Oxford. They were incorporated into a valuable catalogue of the positions of 606 southern stars, for which Mr. Johnson received the gold medal of the Royal Astronomical Society. The instruments belonging to the St. Helena observatory were deposited during many years at the Royal Observatory, Greenwich; they have been since transferred to the Royal Naval Schools, Greenwich Hospital, to be employed in the instruction of the senior boys in the Nautical School.

The principal observatory in the southern hemisphere is that at the Cape of Good Hope. This establishment was founded in the early part of the reign of George IV; an order in council authorising its erection having been signed on October 20, 1820. The building is situated about four miles from Cape Town, and one and a half miles from the beach of Table Bay, at the mouth of the Salt River. The meridian line towards the south passes over Simon's Bay and Cape Point, and towards the north over the east shore of Table Bay and the Blauberg ridge beyond. The centre of Table Mountain lies in a south-westerly direction, about six miles from the observatory. The erection of the building was commenced on September 6th, 1825, and finished in the latter part of 1828. At first the observatory was furnished with the usual astronomical meridional instruments and an equatorial; but the meridional instruments, a transit, and a mural circle, have been since replaced by a new transit-circle, constructed on the same principles and of the same magnitude as the great transit-circle at Greenwich. Observations can therefore be made with the same accuracy as at the parent northern observatory. The Rev. Fearon Fallows was the first astronomer appointed by the Government. He was succeeded in 1831 by Mr. T. Henderson. The present astronomer in charge of the Royal Observatory at the Cape is Sir Thomas Maclear, who has occupied the office since the year 1834, on the appointment of Mr. Henderson to the chair of Practical Astronomy in the University of Edinburgh.

Well equipped permanent observatories have also been established in various other places in the southern hemisphere. They are all in a condition to advance our knowledge of the southern skies. That at Melbourne, under the care of Mr. Ellery, is one of the first class. Australia contains two other public observatories, one at Sydney, the other at Adelaide, besides one or two private establishments. Of the latter, that under the direction of Mr. Tebbutt, of Windsor, New South Wales, deserves a special mention. The principal observatory in South America is that at Santiago, Chili, where a continuous series of astronomical observations has been carried on for several years past.

After the publication of a valuable catalogue of nebulæ in 1833, Sir John Herschel considered that his labours would be incomplete without an examination of the sky of the southern hemisphere in the same systematic manner as that employed by him in his researches in England. He therefore determined at once to transport himself, family, and instruments to the Cape of Good Hope, where he could command that portion of the heavens not visible in Great Britain. On his arrival at the Cape on January 15th, 1834, he took up a temporary residence at Welterfreiden; but soon after, meeting with a suitable mansion at Feldhausen, about six miles from Cape Town, a spot charmingly situated on the last gentle slope at the base of Table Mountain, he selected it for his home during the four years he remained in South Africa. Here he erected, within an inclosure—a kind of orchard, surrounded on all sides by trees—a building for his equatorial, while his twenty-foot reflector was mounted in the open air, due precaution being taken for the protection of the speculum and other delicate parts of the instrument. The results of the labours of the illustrious astronomer have been published in a large quarto volume, full of original observations and research.

As a remembrance of this scientific visit of Sir John Herschel to a distant part of the globe, for no other object than that of a pure love to a science for the advancement of which he has devoted the greater part of his life, a granite obelisk has been erected at Feldhausen, on the site of the twenty-foot reflector, with which most of his observations were made. This obelisk was completed, at an expense of £300, in February, 1842, in the presence of the subscribers, who attended on the occasion of placing the top stone. It stands eighteen feet high from the ground, the base being six feet square and six feet in height. There is an opening on the east face exhibiting the Herschel mark, which points out the site of the twenty-foot reflector. This opening is covered with a bronze plate, containing the inscription of the purpose for which the obelisk was erected.

A proposition has lately been made, and there is every probability of its being carried out, for making a complete survey of the southern heavens, on the same plan as that pursued by M. Argelander, of Bonn, in his survey of the sky of the northern hemisphere. It is intended to observe the exact position of every star, at least, as low as the ninth magnitude, of which there are probably several hundred thousands. Many years must

THE HERSCHEL OBELISK, SOUTH AFRICA.

necessarily elapse before even a major portion of this undertaking can be completed; but by dividing the labour between the principal southern observatories, it is hoped that the work will be so far accelerated as to permit the completion of certain zones of the heavens within a reasonable time. The amount of patience and observing skill required to perform an astronomical work of such magnitude can scarcely be appreciated by any but those who have been more or less accustomed to this class of scientific labour.

NEBULÆ AND CLUSTERS.

To view the general appearance of a nebula, an observer is not required at first to call in the aid of a telescope; for there are several well-known clusters of stars which have for ages been distinguished solely as condensed and bright patches of light. The interesting group of the Pleiades is one of these; the principal stars in Coma Berenices form another woolly group. A somewhat similar cluster of stars called Præsepe, in the constellation Cancer, can be resolved into a number of stars with very slight optical assistance. Some of the principal telescopic nebulæ may even be observed on very clear nights with the naked eye, as the great nebulæ of Orion and Argo Navis. An old chart containing the nebula in Andromeda has been found, by comparing the positions of the stars contained in it with modern determinations, to date as far back as the close of the tenth century. This nebula was thus recognised in the heavens, and its position in relation to the stars was known at least six hundred years before the invention of the telescope.

More than five thousand nebulæ and clusters, including both hemispheres, have been observed and catalogued. Astronomy is indebted for the most part to the two Herschels for the complete analysis of the different varieties of these generally faint objects. Sir William Herschel, in his retired residence at Slough, assisted by his sister Caroline, took advantage of his splendid reflecting telescopes to scour the heavens for these almost invisible particles of star-dust. If anything special in their form was observed, the magnificent forty-foot telescope was directed to them, when many which in smaller instruments appeared as ill-defined matter, were resolved into hundreds or thousands of distinct stars. By a careful and systematic scrutiny, Sir William Herschel separated all objects which are classed under the head of nebulæ into six divisions as follows:—

1. Clusters of stars, in which all the objects are clearly separated, forming one close mass of stars, either in a globular or irregular form.

2. Resolvable nebulæ, or such as give the appearance of being stars, requiring only increased optical power to resolve the whole into distinct stars.

3. Nebulæ where there is no indication of stars, or probability that no telescopic power could alter their appearance. This division has been subdivided into subordinate classes, depending on their magnitude and comparative lustre.

4. Planetary nebulæ. These objects owe their name to an apparent resemblance to a planetary disk. Generally they are either circular, or slightly oval. Some of the planetary nebulæ present some remarkable illustrations of colour. One in the Southern Cross Sir John Herschel noticed to be "a fine and full blue colour, verging somewhat upon green."

5. Stellar nebulæ. This class of objects has generally a condensation of light in the centre of the nebula.

6. Nebulous stars. Here we have a nebulosity with one or more distinct stars shining through it. Nebulous stars are sometimes circular, sometimes oval, sometimes annular, but always of regular form. When the nebula is circular, the star is generally in the centre; when elliptical, two stars are often seen in the foci of the ellipse.

Among the most remarkable of the nebulæ which have attracted the attention of astronomers, the following may be mentioned:—The great nebula in Orion, near Theta Orionis; the great nebula in Argo Navis; the cluster-nebula in Perseus; the nebula in Andromeda; the dumb-bell nebula; spiral nebulæ, etc. The nebulæ in Orion and Argo Navis, and the star-cluster in Perseus, are already briefly described in the notes on these constellations.

The first detailed observations of the nebula in Andromeda were made by Simon

Marius in 1612. It appeared to him "to be composed of rays of light, increasing in brightness as they approached the centre, which was marked by a dull, pale light, resembling the light of a candle seen at some distance shining through horn."

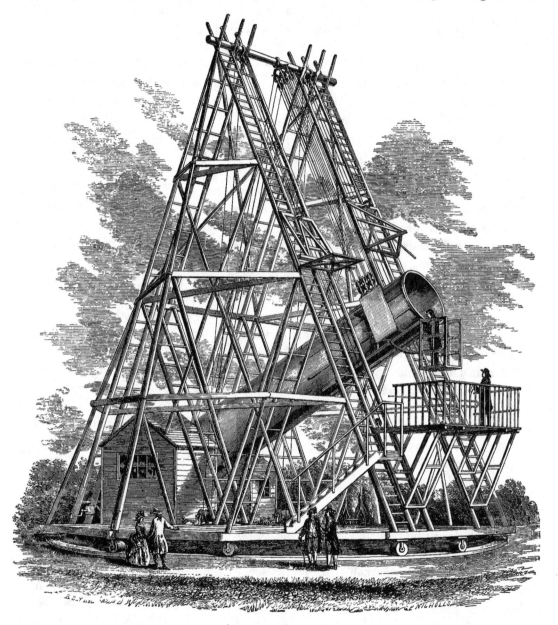

SIR W. HERSCHEL'S FORTY-FOOT REFLECTING TELESCOPE.

According to Sir William Herschel, it is one of the nearest of all the great nebulæ, about one and a half degree in length, and in one of its narrowest places about a quarter of a degree in breadth. The most luminous portion of it approaches the resolvable nebulosity, and exhibits slight indications of a faint red colour. Sir John Herschel, writing in 1826, gives the following detailed description of the great nebula in Andromeda, which he had frequently observed with the great reflecting telescope at Slough. "At present it has not, indeed, a star, or any well-defined disk in its centre; but the brightness, which increases by a regular gradation from the circumference, suddenly acquires a great accession, so as to offer the appearance of a nipple as it were in the middle, of very small diameter, but totally devoid of any distinct outline, so that it is impossible to say precisely where the nucleus ends and the nebula begins. Its nebulosity is of the most perfectly milky absolutely irresolvable kind, without the slightest tendency to that separation into flocculi described in the nebula of Orion, nor is there any sort of appearance of the smallest star in the centre of the nipple. This nebula is oval, very bright, and of great magnitude, and altogether a most magnificent object." Since the date of these remarks, which give the appearance of the nebula as viewed in Herschel's forty-foot reflecting telescope, it has been observed through some of the largest refracting telescopes, especially the fifteen-inch object-glass equatorial of Harvard College, United States, by Mr. G. P. Bond. No signs of resolvability have, however, been detected. The nebula is one of those which give a continuous spectrum like the stars; it is therefore probably only nebulous in an optical sense, owing to the extreme minuteness of its constituent stars. Mr. Huggins remarks, however, that "it may be possible that nebulæ, which have little indication of resolvability, and yet give a *continuous* spectrum, such as the great nebula in Andromeda, are not clusters of suns, but gaseous nebulæ, which, by the gradual loss of heat, or the influence of other forces, have become crowded with more condensed and opaque portions. So far as my observations extend at present, they suggest the opinion that the nebulæ which give a gaseous spectrum are systems possessing a structure, and a relation to the universe, altogether distinct from the great group of cosmical bodies to which our Sun and the fixed stars belong."

The nebula shaped like a dumb-bell or hour-glass is in the constellation Vulpecula. This is another of the irresolvable nebulæ, but it differs so far from that in Andromeda, as to give a spectrum of one bright line only. It is thus evidently composed of gaseous matter. The dumb-bell nebula consists of two luminous symmetrical patches, joined together by a narrow isthmus, the whole of which is surrounded by a faint nebulosity of an oval form. As the appearance of this and most of the other nebulæ changes considerably according to the magnifying power used, it may be remarked that when viewed through a good telescope a low power is always to be preferred when examining these faint objects.

The variety of figure of the nebulæ is very great. They are of every conceivable shape, circular, elliptical, spiral, double, triple, quadruple, and indeed of the most extraordinary irregularity. In many the form changes completely when viewed in telescopes of different magnitudes. A nebula in Taurus, observed by Sir John Herschel, of an oval form of some regularity, when seen in the great reflector of the Earl o   is changed into the form of a gigantic crab, with its full number of legs and claws. A very irregularly shaped nebula in the southern constellation, Dorado, included within the larger Magellanic cloud, is one of the finest objects of its class in the southern sky. The central portion of this beautiful nebula is composed of patches of light and dark alternately, surrounded by a paler nebulosity, over which are scattered a number of very minute stars. The two most remarkable of their regular nebulæ are those near Theta Orionis and Eta Argûs.

The preceding examples of the nebulæ are all taken from the irresolvable class. The clusters which, in low-magnifying telescopes, are seen as nebulous patches of light, are most interesting objects. One in the constellation Hercules is visible to the naked eye on fine nights. It is seen as a luminous round spot, but it may be easily resolved into hundreds of stars, still, however, retaining its globular appearance. Two in the southern sky, one in Centaurus, the other in Toucana, have been sketched by Sir John Herschel. That near Omega Centauri is also visible to the naked eye as a dim round cometic object, in lustre similar to a star of the fourth or fifth magnitude. When seen through large telescopes it is resolved into an immense number of stars from the thirteenth to the fifteenth magnitude, strongly condensed towards the centre. The cluster in Toucana is in the neighbourhood of the lesser Magellanic cloud, in a spot almost free from stars. The condensation in the centre is very marked, and is of a ruddy orange colour. It contrasts very vividly with the white light of the stars immediately surrounding the central part of the cluster.

Clusters, like nebulæ, are to be found scattered over the greater portion of the heavens; they are, however, very scarce in some parts. They are the most numerous in or near the Via Lactea and the Magellanic clouds. The richest region where the globular clusters may be found is certainly in the southern hemisphere, especially in that part of the Milky Way between the constellations Sagittarius, Corona Australis, Scorpio, Ara, and Lupus. The Milky Way itself may be considered as the largest and nearest of the clusters, especially if we include our own Sun as one of its members, as has been speculatively asserted by Sir John Herschel and other astronomers.

NOTES

ON

THE SOLAR SYSTEM.

"Fairest of beings! first created light!
  Prime cause of beauty! for from thee alone
  The sparkling gem, the vegetable race,
  The nobler worlds that live and breathe their charms,
  The lovely hues peculiar to each tribe,
  From thy unfailing source of splendour draw!
  In thy pure shine with transport I survey
  This firmament, and these her rolling worlds,
  Their magnitudes and motions."

THE SUN.

IN the solar system the Sun is the great central source of light and heat. Following the expression of Copernicus, the founder of the modern interpretation of the solar system, it may be termed the light of the world, or according to Theon, of Smyrna, the heart of the universe. Practically, however, it is the centre of a number of independent worlds, non-luminous of themselves, forming, as it were, the focus around which they all revolve. In fact, without its attractive power, all the planets, from Mercury to Neptune, could not be preserved for a moment in their respective orbits, but they would disappear into space, one knows not whither.

As the different planets are separated from each other by a considerable distance, the apparent magnitude of the Sun as seen from each varies in proportion. As viewed from the Earth, the angular diameter of the Sun is rather more than half a degree, so that it would take about 700 Suns, placed side by side, to make a complete circuit of the celestial sphere. If it were possible to transport ourselves to the surface of Mercury, the nearest known planet to the Sun, we should find that the solar diameter would be apparently greatly increased, and that the intensity of the solar light could not probably be borne by inhabitants constituted like ourselves, even for an instant. On Neptune, on the contrary, the Sun would appear only as a brilliant star, the light and heat received at that immense distance being upwards of six thousand times less than that experienced on the surface of Mercury. The actual diameter of the Sun, as determined from the latest astronomical data, is about 856,500 miles, or 108 times greater than that of the Earth. Its superficial extent is 11,679 times greater, while it would require 1,262,153 globes, each of equal magnitude to the Earth, to form a body corresponding in bulk to that of the Sun. But it has been found that the density of the material forming the Sun is about one quarter of the terrestrial density; consequently, if weighed in opposite scales, only 320,000 Earths would be required to equal the total weight of the Sun. These numbers appear large, but they are probably known at the present time within all reasonable accuracy. Whenever any improved determination of the Sun's absolute distance is made, these quantities which represent its bulk must always be corrected in proportion to the alteration of the amount in

215

miles of the distance of the Sun from us, because the radius of the Earth's orbit is the fundamental unit used in the calculation of all the numbers representing the dimensions and bulk of not only the Sun, but of the different members, both primary and secondary, of the solar system. The determination of the value of this fundamental unit in astronomy is considered of so much practical importance, that we do

THE ROYAL OBSERVATORY IN FLAMSTEED'S TIME.

not hesitate to give a brief, but slightly detailed, explanation of the principal methods employed by astronomers to obtain the best possible result.

The Astronomer Royal has stated that the measurement of the Sun's distance has always been considered the noblest problem in astronomy. He says, "It is easy to measure a base-line a few miles long upon this Earth, and easy to make a few geodetic surveys, and easy to infer from them the dimensions of the Earth with great accuracy; and taking these dimensions as a base, common to every subsequent measure, it is easy to measure the distance of the Moon with trifling uncertainty. But the measure of the Moon's distance in no degree aids in the measure of the Sun's distance, which must be undertaken as a totally independent operation. A second reason is that, in whatever way we attack the problem, it will require all our care and all our

THE ROYAL OBSERVATORY, GREENWICH.

ingenuity, as well as the application of almost all our knowledge of the antecedent facts of astronomy, to give the smallest chance of an accurate result. A third reason is, that upon this measure depends every measure in astronomy beyond the Moon; the distance and dimensions of the Sun and every planet and satellite, and the distances of those stars whose parallaxes are approximately known." Although, however, this problem is one of the most attractive in astronomy, yet its practical solution is a very difficult matter indeed : in proof of which it may be stated that several of the most eminent astronomers, from the days of Hipparchus down to the present time, have devoted much original thought on this subject, and yet it cannot be said that an absolutely correct result has been obtained, although there is very little doubt that we now know the true distance of the Sun within very small limits.

There are several methods by which the distance of the Sun may be determined. Some of them are, however, of too abstruse a nature to be even mentioned here. That which is considered to be the best method is by the observation of the transit of the planet Venus across the solar disk at different stations on the Earth's surface. This phenomenon occurs only at long intervals, the last having taken place in 1769, while the next will be in 1874. When viewed from stations widely separated in latitude, the portion of the disk of the Sun traversed by the planet is apparently very different in two opposite localities, arising from the effects produced by the displacement of the true path of the planet on account of parallax.* By observing the accurate local time of the ingress and egress at the different stations, the exact time occupied by the planet in passing from one edge of the Sun to the opposite edge defines this apparent 'path of Venus over the solar disk, and gives the relative displacements caused by the different amounts of parallax, as viewed at the respective stations. From these observed data, the horizontal parallax of Venus can be determined, and from it the distance of Venus in miles from the Earth. We know already, with the greatest exactness, by means of Kepler's third law, that when the mean distance of the Earth from the Sun is represented as unity, that of Venus is 0·723 ; consequently, by the simple rule of proportion, it is a very easy matter to deduce indirectly the distance of the Sun when that of Venus is known; or, in like manner, of any other member of the solar system.

Although, however, the mere observation of the amounts of the displacements of a planet on the solar disk as seen from two widely-separated stations is no very difficult matter when we can get the opportunity, yet in consequence of the very minute angular value of solar parallax—less than nine seconds of arc—and the extreme care required to observe the phenomenon accurately, a large number of observations by different observers are necessary before a resulting parallax can be deduced correct within a few hundred parts of a second of arc. This preciseness must be obtained, or the labour and expense of the expeditions are utterly useless. But our readers may naturally ask, What is meant by the angle of a second of arc? This quantity

---

* *Parallax* is the apparent change in the position of an object due to a change in the position of the observer. In celestial objects, it is the angle under which a line drawn from the station of the observer to the centre of the Earth would appear at the object observed. Parallax is always the greatest when the object is in the horizon, where it is called the horizontal parallax, diminishing gradually to the zenith, where it is nothing. It is only appreciable in the positions of Sun, Moon, planets, and comets, the fixed stars being far too distant to be sensibly affected by it, even if the station of the observer be changed from the north to the south pole. In a few cases, however, the effects of parallax have been detected in the positions of some of the nearest of the fixed stars when viewed from opposite sides of the Earth's orbit.

is so minute that the unassisted eye is totally unable to appreciate it. Mr. Pritchard remarks:—"It will convey but little idea if we say it is the 324-thousandth of a right angle, for the very numbers confuse the mind. But what then is a second? It is equivalent to the angle subtended by a ring one inch in diameter, viewed at the distance of three miles and a third. The correction to be made to the Sun's parallax is just one third of this; that is to say, it is the error which a rifleman would make who shot at the right-hand edge of a sovereign placed *twelve* miles off, and who hit it by mischance just on the left edge! It is what a human hair would appear to be, if viewed at the distance of above 150 feet! Such are the quantities with which astronomy, of necessity, deals." If, then, a second of arc is so minute a measurement, what must we say when this second is again subdivided into tenth or hundredth parts, every one of these hundredth parts of a second of the solar parallax representing 100,000 miles in the determination of the distance of the Sun from the Earth? And yet this almost mathematical accuracy is obtained in such observations as the transit of Venus over the Sun's disk, if the observers are favoured with a serene sky, at those stations where both the times of the ingress and egress of the planet are successfully observed.

The advantage of determining the distance of the Sun by means of the observations of the transit of Venus over the solar disk, was first pointed out by Dr. Halley, in the year 1716, and again a few years before his death. Although it was impossible for himself to see any practical result arising from his suggestions, he clearly demonstrated the possibility of observing the transits which were to occur in the years 1761 and 1769, remarking that he bequeathed the problem of ascertaining the distance of the Sun from the Earth, according to a method proposed by him, as a task to posterity. The principal astronomers of the last century gladly accepted the bequest, for extensive arrangements were made by the leading astronomers of the day to carry out Dr. Halley's views, first in the year 1761, and secondly in 1769. In the former year, some of the most experienced astronomical observers of Europe were sent to various parts of the world, but from cloudy weather and other unfavourable causes, the results obtained were not so accordant as was desired. Partial failure in the first instance only stimulated the observers to make renewed exertions in the preparations for the next transit in 1769. Expeditions were sent to distant places for this object, principally at the expense of the Governments of several of the European nations. The English Government gave a very material assistance, by fitting out an exploring expedition to the South Seas under the command of the celebrated Captain James Cook. One of the principal instructions given to this great navigator was to select a suitable position in the island of Otaheite for the purpose of observing the times of the ingress and egress of the planet Venus at its transit over the solar disk on June 3, 1769. Mr. Green, formerly an assistant at the Royal Observatory, accompanied the expedition as astronomer. Observations of both ingress and egress were made by Captain Cook, Mr. Green, and Dr. Solander, at a spot which is known as Venus Point to this day.

Messrs. Wales and Dymond, both formerly of the Royal Observatory, made similar observations at Fort Prince of Wales, Hudson's Bay; MM. l'Abbé Chappe, Doz, Pauly, and Medina, at St. Joseph, in California; M. Rumowski, at Kola, in Lapland; and Fathers Hell and Sajnovics and M. Borgrewing, at Wardoë, an island in the Arctic Ocean, at the north-eastern extremity of Norway. As soon as the observations were received in Europe, they were subjected by some of the principal mathematicians to a severe critical discussion. M. Lalande published his investigations as a separate treatise. The results on the whole were considered to be tolerably accordant, although the greatest and least differed 0″·39 in the value of the "horizontal equatorial solar parallax," equal to four millions of miles in the Sun's distance. The separate values of the resulting solar parallax were according to

		″
Euler	. . . .	8·82
Pingré	. . . .	8·81
Hornsby	. . . . . .	8·78
Maskelyne	. . . . .	8·72
Lexell	. . . .	8·68
Smith	. . . .	8·61
Lalande	. . . . . .	8·50
Planmann	. . . . .	8·43

The preceding results were not obtained from the observations of the observers previously mentioned alone, but partially from those of other observers stationed in various localities where only one phase was visible, the ingress in some places, and the egress in others. In 1822 and 1824, in consequence of the advancement of astronomical knowledge, M. Encke, a celebrated German astronomer, published two elaborate treatises on the subject, the first containing an ample scrutiny of the observations of 1761, and the second a thorough critical investigation of those of the transit of 1769. Using the improved data at his command, he gave as the value of mean solar parallax the definitive result of 8″·5776, which was at once accepted by astronomers as the best data in existence for the determination of the Sun's distance. Hence for many years the numbers 95,300,000 miles which were deduced from it came into general use. A result very similar to Encke's was previously obtained by M. de Ferrer.

This new value of the Sun's distance was received with so much confidence as to become almost a "household word" in astronomy, for a period of nearly forty years. The first doubt of its absolute truth came from Professor Hansen, of Gotha, who, while constructing new lunar tables, found that to satisfy the more refined observations of the Moon now made at Greenwich, it was necessary to make a considerable increase in the value of the solar parallax. M. Le Verrier, also, while preparing new tables of the movements of Venus, the Earth, and Mars, announced that to make the theory agree with the observations, the solar parallax required an increase of one-thirtieth part. Attention being thus drawn to this subject, the Astronomer Royal gave, in 1857, an oral address before the members of the Royal Astronomical Society, in which he strongly urged the necessity of making early preliminary arrangements for the proper observation of the next transits of Venus, which will take place in the years 1874 and 1882. But at the same time he pointed out that when the planet Mars is in or at its nearest distance from the Earth, the observation of its parallax would give indirectly a value of the Sun's distance little inferior to that obtained from the transit of Venus. In 1862, therefore, an extensive series of observations was made, with first-class fixed instruments, at Greenwich, Pulkowa, Washington, and Albany, in the northern hemisphere, and at the Cape of Good Hope, Williamstown, near Melbourne, and Santiago, in Chili, in the southern hemisphere. These observations have been investigated by different astronomers independently, the result of each confirming in a remarkable manner the theoretical deductions of MM. Hansen and Le Verrier.

The absolute agreement in the mean values of solar parallax, determined by methods totally different in their manner of solution, pointed to the necessity for a re-discussion of the observations of the transit of Venus made in 1769. M. Powalky was the first to do this, but although the result obtained by him agreed nearly with the increased value shown by the recent determinations, yet the method adopted in his calculations was open to several objections. Mr. Stone has, however, made a subsequent investigation, by interpreting the observations strictly as recorded in the notes of the observers, and by dividing the phenomenon into the two phases of real and apparent internal contact caused by the effects of irradiation produced by the intense light of the Sun. The interval between the two phases was found to be about nineteen seconds of time. M. Encke, on the contrary, treated all the observations as real internal contacts, concluding that where any disagreement between the recorded times of the different observers at the same station took place, the discordance probably arose from a personal error made by the observer. The interpretation given by Mr. Stone is undoubtedly the correct one, an opinion which has been endorsed by the principal English astronomers. Thus after the lapse of a century since the transit occurred, the originality of one mind has unravelled a mystery which had perplexed the lovers of astronomy since M. Hansen first expressed his opinion in November, 1854, as to the necessity of a considerable increase in the value of the solar parallax. According to Mr. Stone's researches, the true value given by the observations of 1769, is $8''\cdot91$, agreeing sensibly with those determined by M. Hansen, M. Le Verrier, from the observations of Mars in 1862, and from the experiments of M. Foucault on the velocity of solar light. In some remarks on this subject, the Rev. Charles Pritchard says:— "Thus all now has become clear in this very intricate question. We will not say thus has been removed the opprobrium from astronomy, for to astronomy it was never in reality an opprobrium. The physical circumstances attending the passage of a dark body over a very bright one, and then viewed through a telescope, were not understood at the time when the observations were made, and it was these which produced, not the astronomical error, but the then inextricable difficulties of the case. The error arose from the observers of the transit seeing without perceiving, and it has been most successfully removed by Mr. Stone, who perceived the meaning of the phenomena without seeing them." The mean, or average, of all the trustworthy recent determinations is 8″·90. This value of the solar parallax, which is the angular measure of the Earth's equatorial semidiameter viewed at the mean distance of the Sun, gives for that mean distance, with only a small probable error, 91,841,000 miles.

If we view the surface of the Sun with an ordinary astronomical telescope, numerous small black irregular patches or spots are generally visible. Sometimes as many as eighty have been counted at one time, while at periods near the minimum frequency, the solar disk is perfectly free from them. Occasionally these spots are of sufficient magnitude to be observed with the unassisted eye, protected only by a coloured glass. Around the principal spots there is generally a fringe of less density called the penumbra. After a spot has been under examination for a short time, it may frequently be observed to undergo considerable alterations of form, while the central part, or nucleus, which is generally much darker than the rest, has been seen on many occasions to revolve. A remarkable instance of rotatory motion was noticed by the Rev. W. R. Dawes, in January, 1852, when the whole spot was observed to have rotated in six days completely around the small black nucleus. The variations in the appearance of the spots do not follow any rule, for occasionally a spot will keep its general contour from the time it enters on the solar disk to the moment of its disappearing on the opposite side, while others have been known to break up within a day or two into twenty or thirty distinct portions. Some extraordinary variations of this kind can be seen in Mr. Carrington's records of solar spots; and the Rev. F. Howlett has laid before the members of the Royal Astronomical Society special accounts of two remarkable spots, the first visible from July 25th to August 4th, 1862, the second in October, 1865. The dimensions of the latter extended to 110″ in length, and 60″ in breadth; and making allowance for its irregularity of form and density, it spread over a superficial area of not less than 972 millions of square miles. Adding to this, the displacements caused by numerous small spots in other directions, the solar photosphere was disturbed to the enormous extent of 1,137,000,000 square miles, or six times that of the whole surface of the Earth! Solar spots have, however, been observed far exceeding in magnitude those recorded by Mr. Howlett. In June, 1843, a very remarkable one was visible without the aid of a telescope during a whole week; its diameter, according to M. Schwabe, being about 77,000 miles, or nearly ten times that of the Earth. On May 25th, 1837, Sir John Herschel, while located at Feldhausen, Cape of Good Hope, noticed a spot of enormous size, the black centre, or nucleus, being large enough to have allowed a globe of similar magnitude to the Earth to drop through it, leaving a thousand miles clear of contact on all sides of the brink of the tremendous gulf.

Owing to the rotation of the Sun on its axis, which occupies about 25^d 8^h, solar spots can be seen to apparently traverse the Sun's disk from east to west. Some which have disappeared at the western edge have been again recognised on their reappearance at the eastern edge, although their form may have undergone considerable modification during the interval. The spots are entirely confined within an equatorial zone, extending between thirty and forty degrees each side of the solar equator, or rather between two zones, one north and the other south of the equator, the intermediate band on the equator itself being comparatively free from them. They are never seen near the poles. Many speculative theories on the nature of solar spots have been put forth by several leading astronomers, but their probable origin is still open to discussion.

The continuous mapping down of the Sun-spots from day to day, persistently carried on for years by M. Schwabe, of Dessau, by Mr. Carrington, and at the Kew Observatory, has produced an accumulation of most valuable facts relating to the distribution of these interruptions to the uniform brilliancy of the solar photosphere. It has been discovered, principally from these records, that the frequency or non-frequency of solar spots is not altogether an accidental circumstance, but that there is a considerable regularity in the times of maximum and minimum. From the observations of M. Schwabe, it seemed tolerably clear that the interval between maximum and maximum generally consisted of about ten years. At one time this periodical variation was supposed to be connected with a similar periodical change in terrestrial magnetic declination, or the variation of the compass. Later observations have, however, thrown some doubt on the apparent coincidence. The positions of Venus and Jupiter in certain parts of their orbits have also been noticed to coincide with the maximum frequency of Sun-spots; this connection, however, is at present far from proved, and a much more extended series of observations must be made before any results deduced from the possible effect of planetary influence on the solar photosphere will be considered as satisfactorily established.

Dark spots of irregular form are not the only objects seen on the solar surface when viewed through an astronomical telescope. It is not very difficult to detect near the limbs, or edges, of the Sun, and also near ordinary Sun-spots, some very luminous streaks of much greater relative lustre than the surrounding photosphere. These streaks are known by the name of *faculæ*. They are generally many thousands of miles in length, and frequently connected one with the other over a space equal to thirty or forty thousand miles in length, and from one to four thousand miles in breadth. Other singular appearances on the solar disk are distinctly visible when seen through large telescopes. These phenomena have been distinguished principally by the names of granules, willow-leaves, rice-grains, etc. Observations of the general granular aspect of the solar surface have been frequently made, but the first detection of the small interlacing particles, termed willow-leaves, which are quite distinct things form the ordinary granules, was made by Mr. James Nasmyth, the inventor of the steam-hammer. A considerable controversy was carried on for some years as to the real existence of these objects, but they have now been seen by so many leading

223

astronomical observers that little doubt remains that these apparently minute particles form an important part of the photosphere. The first announcement by Mr. Nasmyth was made in a private letter to a friend, and afterwards communicated to the public. Mr. Nasmyth remarks that "the filaments in question are seen, and appear well-defined, at the edges of the luminous surface where it overhangs the penumbra, as also in the details of the penumbra itself, and most especially are they seen clearly defined in the details of 'the bridges,' as I term those bright streaks which are so frequently seen stretching across from side to side over the dark part of the spot. So far as I have as yet had an opportunity of estimating their actual magnitude, their average length appears to be about 1,000 miles, the width about 100. There appears no definite or symmetrical arrangement in the manner in which they are scattered over the surface of the Sun, for they appear across each other in all possible variety of directions."

Among the numerous observers who have seen these well-defined solar particles, we may mention M. Secchi, whose observations made at Rome on April 11–13, 1869, are peculiarly interesting. He noticed that the penumbra of a solar spot under examination was covered with the small luminous particles, termed "willow-leaves," all of which were nearly identical in form. They were arranged apparently in bundles, converging towards the centre of the spot, across which they were thrown in the form of a bridge, arranged in a double line one after the other. M. Secchi considers that the "willow-leaves" form a kind of net-work foundation of the photosphere. The observations of M. Secchi represent very closely what has been seen by the aid of the great thirteen-inch object-glass equatorial at the Royal Observatory, Greenwich.

By his spectroscopic researches on the solar spots, M. Secchi has proved satisfactorily the evident hollow structure of these phenomena. Moreover, he has observed that the cavities are filled with dense metallic vapours, such as may be supposed to exist at the bottom of the solar photosphere. This is only one of the numerous solar discoveries made since the application of spectrum analysis to the comparison of solar light with that emitted by terrestrial substances heated to a state of incandescence, by which our knowledge of the principal elements contained in the solar photosphere has been so considerably increased.

Since the discovery in August, 1868, of the gaseous nature of the rose-coloured prominences seen in all total eclipses of the Sun, several astronomers, including M. Janssen, Mr. Lockyer, Mr. Huggins, and M. Secchi, have clearly demonstrated that, in addition to the ordinary photosphere, a narrow belt of coloured gaseous matter—principally hydrogen in a state of combustion—extends over the whole surface of the Sun. Hitherto, portions only of this continuous bed of self-luminous matter have been observed during total eclipses of the Sun, the irregularities or detached portions being known as the rose-coloured protuberances. A large part of the Moon's edge was seen fringed with this rose-coloured matter, during the eclipse of 1851, by the Astronomer Royal, Mr. Hind, Mr. Dawes, and others. In the eclipse of 1860, M. Le Verrier noticed it also. "The Sun," he remarks, "is simply a luminous body on account of its high temperature, covered by a continuous bed of rose-coloured matter, the existence of which is undoubtedly proved." M. Janssen's and Mr. Lockyer's independent

THE GREAT EQUATORIAL, ROYAL OBSERVATORY, GREENWICH.

spectroscopic experiments since the solar eclipse of August 18th, 1868, have definitely confirmed what was before considered mixed up with speculation. The first observation in sunlight of the rose-coloured matter composing the protuberances, is very remarkable.

Y

M. Janssen was so struck with the brilliancy of the bright lines in the spectrum of the protuberances during the eclipse, that he resolved to continue his observations after the totality was over, if the weather remained favourable. Unfortunately, the Sun disappeared directly after the eclipse, and continued obscured during the remainder of the day. However, soon after sunrise on the next morning, M. Janssen succeeded far beyond his expectations, for he was not only enabled to observe the bright lines of a protuberance, but he could easily infer from them the rapid changes constantly going on in the form of the prominence. Mr. Lockyer first saw the bright lines in sunlight on October 20th, 1868, before he was aware of the success of M. Janssen in India, which was not announced in Europe before October 26th. Each observer may therefore justly claim a share in this important discovery; but to M. Janssen undoubtedly belongs the honour of being the first person who observed these curious appendages of the Sun without the intervention of the Moon in a total eclipse.

The observation of the rose-coloured matter around the edge of the solar disk has been very successful, not only by M. Janssen and Mr. Lockyer, but by Mr. Huggins, M. Zöllner, and Padre Secchi. Professor Zöllner, of Leipsic, has published some remarkable delineations of the rose-coloured protuberances as observed by him on July 1st and 4th, 1869. The rapidity of motion of the composing matter is very clearly shown in the different sketches of the same prominence, some of them having been made at intervals of only a few minutes of time. Mr. Lockyer has on frequent occasions noticed these rapid movements of the gaseous solar matter, sometimes drifting with a velocity of forty miles a second in a vertical direction, and one hundred and twenty miles in a horizontal direction.

It is useless to speculate at present on what insight we may ultimately obtain into the chemical composition of our great central luminary, but the systematic examination of the solar disk, by means of the powerful spectroscopes now in the hands of some of our most eminent amateur astronomers, cannot fail to increase our still imperfect knowledge of solar physics to an extent of which at present we can have only a vague conception.

TOTAL ECLIPSES OF THE SUN.

"As when the Sun new risen
Looks through the horizontal misty air.
Shorn of his beams, or from behind the Moon
In dim eclipse disastrous twilight sheds
On half the nations, and with fear of change
Perplexes monarchs : darkened so, yet shone
Above them all th' Archangel."—*Paradise Lost.*

THE most ancient record of a solar eclipse is to be found in the *Shoo King*, the oldest historical treatise in the Chinese language. In that work it is related that in the reign of Chung Kang, the fourth emperor of the Hea dynasty, an eclipse of the Sun took place. The passage recording this event occurs incidentally in an account of He and Ho, two important scientific officers, who appear to have had the superintendence of astronomical questions, and were responsible to the emperor for the due announcement of all predictions of astronomical phenomena, which were always required, on account of certain religious ceremonies having to be performed when anything unusual took place. Being, however, rather too fond of wine, these unfortunate astronomers neglected their duties, and consequently, failing in the prediction of the eclipse in question, rendered themselves liable to the punishment of death, which, according to the old chronicle, was duly inflicted. In the *She Ke*, a collection of Chinese history, in three hundred volumes, this tragic anecdote is repeated. Mr. Williams, of the Royal Astronomical Society, after a critical examination of the *Shoo King*, and other Chinese works, has come to the conclusion that this eclipse occurred in the year B.C. 2158.

The record of several eclipses of the Sun has been handed down to us in the works of the Greek and Roman historians. Most of them have been mentioned in connection with some special historical event. Accurate epochs for chronological reference have been obtained by the calculation of the dates of these eclipses by means of the refined astronomical tables of the Sun and Moon now in use. The eclipse of Thales of Miletus, as recorded by Herodotus, is one of the most important of these ancient eclipses, the exact date of which has been determined. No chronologer considers this subject of greater importance than the present Astronomer Royal, to whom we are indebted for the definitive settlement of the date of this old historical epoch. Using in his calculations the new lunar tables of Professor Hansen, which were constructed from the Greenwich observations of the Moon, Mr. Airy has finally decided that the eclipse must have taken place on May 28th, B.C. 585. The substance of the account of the battle fought on the day of this eclipse, as related by Herodotus, is—that upon the refusal of Alyattes, king of the Lydians, to deliver up some Scythian fugitives to Cyaxares, king of the Medes, war broke out between the two nations. Hostilities continued between them during five years with equal success. In a battle which took place in the sixth year of the war, the day suddenly turned into night. The effect on the soldiers through this unlooked-for transition from light to darkness was so great, that, probably from fear, both sides ceased fighting simultaneously, and each endeavoured anxiously to obtain peace. This eclipse was foretold by Thales some time before the battle, and is the most ancient record of an undoubted astronomical prediction.

Xenophon relates in his *Anabasis* that when the Persians obtained the empire of the East from the Medes, the king of the Persians besieged the city of Larissa, supposed by Layard, Jones, and others to be the modern Nimroud, but he could not succeed in capturing it. But a cloud covered the Sun and caused it to disappear completely, creating so great a consternation among the inhabitants that they withdrew; and thus the city was taken. It has been found, from calculation, that a total eclipse of the Sun actually occurred on May 19th, B.C. 557, a date which agrees sensibly with that on which the disappearance of the Sun is related to have taken place at Larissa. As a further proof of the identity of the ancient record with modern research, it has been found that the central band of totality passed over the district known now as Turkey in Asia.

A phenomenon, which appears to have been very similar to a total eclipse of the Sun, is recorded to have taken place about the year B.C. 480. It is thus described by Herodotus:—"With spring the army, being ready, set out from Sardes on its march to Abydos; and as it was setting out, the Sun, leaving his seat in heaven, became invisible, when there were no clouds but a perfectly clear sky; and instead of day it became night. Xerxes, who saw this and heard about it, felt some anxiety, and inquired of the magi what the appearance portended; they replied that the deity prognosticated to the Greeks the desertion of their cities; saying that the Sun was the prognosticator for the Greeks, the Moon for the Persians. When Xerxes heard this he was very joyful, and proceeded on his march." This event, which has been interpreted as a total eclipse of the Sun, has given much trouble to chronologers, who cannot make the historical record agree with modern calculations.

Another very important chronological eclipse is recorded by Diodorus Siculus. The fact of this solar eclipse having occurred, as related by the historian, has been confirmed beyond question by the Astronomer Royal and others. The computed date is

August 15th, B.C. 310. Agathocles, being blockaded in the harbour of Syracuse by the Carthaginians, secretly formed the design of invading the Carthaginian territories, but was unable for several days to evade the enemy's fleet. At length a convoy of provision-ships appeared; the blockading ships left their stations to attack the convoy, when Agathocles seized that opportunity for leaving the harbour. The Carthaginians soon discovered that Agathocles had fled, when they left the convoy and followed him, but he escaped with difficulty under cover of the night. "The next day," says Diodorus, "there was such an eclipse of the Sun that the day wholly put on the appearance of night, stars being seen everywhere."

A well-authenticated total eclipse of the Sun occurred on August 31st, A.D. 1030. It is well known in Norwegian history as causing a great consternation at the battle of Sticklastad, during which the Danish king Olaf was slain.

In addition to the preceding ancient records of total eclipses, many others have taken place, especially in China, but very few reliable data can be gathered historically concerning them, though the truth of several has been confirmed by modern researches.

Phenomena of this kind are of rare occurrence in any one particular portion of our globe. In England, for example, no total eclipse of the Sun has been observed since the 3rd May, 1715, new style. On this occasion the narrow band of total phase passed over London. Several other total eclipses have been recorded by some of our ancient chroniclers as having taken place in former times. In an old Saxon chronicle relating to the events of the year 1140, it is recorded that "In the Lent the Sun and the day darkened about the noon-tide of the day, when men were eating; and they lighted candles to eat by. That was the thirteenth day before the calends of April. Men were very much struck with wonder." William of Malmesbury also states, in reference to this eclipse, that while persons were sitting at their meals, the darkness became so great that many thought that chaos was coming, and, on going out, several stars were seen near the Sun. According to the computations of Dr. Halley, this eclipse took place in March, 1140. A remarkable total eclipse of the Sun, visible in Scotland, occurred on June 17th, 1433. The time of complete obscuration was long remembered in that country as the *Black Hour*. Another took place in 1598, on a Saturday, which was frequently alluded to for a considerable period in the border counties of England and Scotland, as *Black Saturday*.

When numerous objects are scattered over the firmament at different distances and in varying positions, it sometimes happens that three objects will come into the same line, or nearly so, by the interposition of the central object. Now this is precisely the case in an eclipse of the Sun, when we have the Sun, Moon, and Earth in a line, and, as viewed from the surface of the Earth, the Moon is the intermediate body. In a total eclipse of the Sun it happens that, at the greatest phase, the centre of the Moon passes over the centre of the Sun, and as on these occasions the apparent diameter of the Moon is slightly in excess of that of the Sun, the latter body becomes invisible to the inhabitants of a small portion of the Earth contained within the zone of totality. In favourable eclipses, the duration of total darkness varies from three to six or seven minutes. This, however, depends on the position of the Moon in its orbit, and in the heavens, at the time of total darkness; because its apparent diameter visibly increases or diminishes from day to day as it approaches or recedes from the Earth. The augmentation of the Moon's diameter is also very sensible when at its greatest altitude. The Sun's apparent diameter also varies in the course of the year, but in a less degree. For an eclipse to take place with the maximum period of total darkness, the three following conditions should be fulfilled. 1.—The Sun must be at its greatest distance from the Earth, so that its angular diameter would be at the minimum. 2.—The Moon, on the contrary, must be at its least distance from the Earth, so that its angular diameter would appear at the maximum. 3.—The position of the Sun and the Moon in the firmament at the moment of total darkness should be in or near the zenith of the place of observation. These three conditions were very nearly fulfilled in the great solar eclipse of August 17th–18th, 1868, in which the total darkness, at stations near the Gulf of Siam, continued during six minutes and fifty seconds. When at the time of an eclipse the Moon is in apogee, or at its greatest distance from the Earth, its apparent diameter is less than that of the Sun, so that although the two centres are coincident as before, no total eclipse can take place; but the Sun being now the apparent larger body, a ring is formed around the dark Moon, causing an annular eclipse. One of this kind was very satisfactorily observed in the north of England on Sunday, May 15th, 1836, and attracted considerable attention throughout the British Isles. When the centres of the Sun and Moon are not coincident, as viewed from the Earth, we have then a partial eclipse. This last phenomenon is of frequent occurrence.

The principal phenomena visible during a total eclipse of the Sun are the following:— 1.—The corona, or ring of light, surrounding the dark body of the Moon. 2.—The brilliant star-like points seen immediately before the commencement or after the end of complete obscuration, commonly called "Baily's beads." 3.—Irregular flame-looking protuberances on the dark edge of the Moon, usually of a pink or rose colour. 4.—Effects of total obscuration on scenery and animals. Our remarks will therefore be confined briefly to these four phenomena.

The luminous ring, or corona, is an exceedingly interesting object. Ancient, as well as modern, philosophers have written of its beautiful appearance. Philostratus, in his *Life of Apollonius*, says, " In the heavens there appeared a prodigy of this nature. A certain corona, resembling the iris, surrounded the orb of the Sun and obscured his light." Plutarch has also alluded to the corona. Confining ourselves to more modern observations, Dr. Halley, who observed the eclipse of May 3rd, 1715, from the apartments of the Royal Society, Crane Court, Fleet Street, remarked that " A few seconds before the Sun was all hid, there was discovered around the Moon a luminous ring, about a digit, or perhaps a tenth part of the

Moon's diameter in breadth. It was of a pale whiteness, or rather pearl colour. . . . . The ring appeared much brighter and whiter near the body of the Moon than at a distance from it." M. De Louville, a French *savant* who observed the eclipse with Dr. Halley, also noticed the luminous ring, which appeared to him coloured with a deep red around the edge of the Moon. Mr. Baily, who observed the eclipse of July 8th, 1842, at Pavia, remarked that the breadth of the corona, if measured from the circumference of the Moon, appeared to be nearly equal to half the Moon's diameter. It had the appearance of brilliant rays, the colour being quite white. The rays had a vivid and flickering appearance, something similar to that which a gaslight illumination might be supposed to assume. On the same occasion Mr. Airy saw the eclipse from the neighbourhood of Turin, when he noticed the corona as a ring of faint, nearly white light; but clouds interfered in some measure with his observations. In the eclipse of the 28th July, 1851, the corona was seen by Mr. Airy much more favourably. He remarks that "The corona was far broader than that which I saw in 1842. Roughly speaking, its breadth was little less than the Moon's diameter, but its outline was very irregular. I did not remark any beams projecting from it which deserved notice as much more conspicuous than the others, but the whole was beamy, radiated in structure, and terminated (though very indefinitely) in a way which reminded me of the ornament frequently placed round a mariner's compass. Its colour was white, or resembling that of Venus." Appearances similar to those described have been seen at different eclipses by the numerous observers who have now witnessed a total eclipse. It is not necessary, however, to examine the separate accounts of these astronomers, whose notices of the corona agree sensibly with those of Mr. Airy and Mr. Baily.

The origin of the corona has been the subject of many astronomical discussions. It was suggested by Kepler, supposing the phenomenon were connected with the Moon, that an explanation of it could be given by conjecturing that, as the rays of light were proceeding from the Sun to the Earth, they might be refracted whilst penetrating the Moon's atmosphere, occasioning an appearance analogous to the corona. Dr. Halley looked with some favour upon this hypothesis. Because this astronomer, in the eclipse of 1715, observed that the corona appeared to him to be concentric with the Moon, he concluded that it was possibly owing to the Moon's atmosphere, as suggested by Kepler. Dr. Halley does not, however, give this opinion without some qualification; for he has stated that the great length of the rays far exceeded the height of the Earth's atmosphere; and " the observations of some who found the breadth of the ring to increase on the west side of the Moon as the emersion approached, together with the contrary sentiments of those whose judgments I shall always revere, make me less confident, especially in a matter whereto, I must confess, I gave not all the attention requisite." In the opinion of Dr. Halley, the corona was much brighter and whiter near the edge of the Moon than at a distance from it. It resembled, in all respects, the appearance of an enlightened atmosphere viewed

231

from a great distance: but he would not undertake to decide finally from the observations, to which body the luminous ring belonged. With regard to the intrinsic brightness of the corona, most astronomers have seen it disappear at the appearance of the first direct ray of light from the Sun; but M. Secchi, in 1860, distinctly saw the corona forty seconds after the end of total obscuration.

Various experiments have been made to obtain an artificial illustration of the corona by M. De La Hire, M. Delisle, Professor Powell, and others. No very decided result, however, has been obtained from their researches.

So much attention has been given to this phenomenon at every eclipse since 1842, that most astronomers have now agreed to consider that the corona is an appendage of the Sun, although the matter of which it is composed is still a mystery. In the eclipse of August, 1868, Captain Branfill observed that the corona was strongly polarised everywhere in a plane passing through the centre of the Sun.

"Baily's beads," or the breaking-up of the narrow line of the Sun into brilliant particles immediately before total darkness, have not been seen by some eclipse observers; while, on the contrary, others have noticed and described them as a phenomenon of marvellous beauty. The immediate cause of the beads has on this account been the subject of much discussion, and even now cannot be said to have received a final settlement. The most probable explanation is, that the solar rays are transmitted through the valleys on the Moon's edge, while they are at the same time cut off by the mountainous ridges. These transmitted rays being affected by irradiation, or by the false light produced by their passage through the varying densities of the different strata of the Earth's atmosphere, or from optical defects in the telescope, give that sparkling appearance which has been very truthfully compared to a diamond necklace. The author saw an effect precisely similar to this at Christiania, during the eclipse of July 28th, 1851; for, while anxiously watching for the total obscuration, which took place in fifteen seconds afterwards, he saw the narrow line of the Sun suddenly break into these small particles of light. They were of different magnitudes, some of the beads being merely brilliant points, while others were larger and elongated in form. He could not help exclaiming to his assistant, who was counting a chronometer at his side, "O what a glorious sight; here is a necklace of most brilliant diamonds!" In fact, he was unprepared for so magnificent a sight, although he was expecting something very startling. At the reappearance of the Sun the same general phenomena were exhibited; the effect on the imagination was not, however, so striking, but there was no sensible difference in the brilliancy of the beads.

This phenomenon, though distinguished by the name of Mr. Baily, had been noticed by many previous observers. Dr. Halley remarks that, in 1715, "about two minutes before the total immersion, the remaining part of the Sun was reduced to a very fine horn, whose extremities seemed to lose their acuteness, and to become round like stars; and, for the space of a quarter of a minute, a small piece of the southern horn seemed to be cut off from the rest by a good interval, and appeared like an oblong star." M. Nicolai, who observed at Mannheim the annular eclipse of September 7th, 1820, states that, about a second before the annulus was formed, the fine curve of the Moon's disk, then immediately in contact with the edge of the Sun, appeared broken into several parts; and, in a moment, these parts flowed together like drops of water or quicksilver near each other. M. De Zach saw similar phenomena during the same eclipse. Mr. Baily has given a graphic account of the beads seen by him in the great annular eclipse of Sunday, May 15th, 1836, which has been the means of identifying his name with them from that time. "My surprise was great," he remarks, "on finding that these luminous points, as well as the dark intervening spaces, increased in magnitude, some of the contiguous ones appearing to run into each other, like drops of water; for the rapidity of the change was so great, and the singularity of the appearance so fascinating and attractive, that the mind was for a moment distracted and lost in the contemplation of the scene, so as to be unable to attend to every minute occurrence."

Interesting as the preceding objects are to the telescopic observer, they have only a secondary rank in comparison with those extraordinary and mysterious appendages seen only during totality, and known as "red prominences." These consist of protuberances, generally with a more or less reddish tinge, scattered promiscuously on the edge of the Moon, and projected on to the white corona. Some of these prominences are of colossal size, reaching to a height of fifty or sixty thousand miles or more: the majority, however, are much smaller. Some are seen detached, and apparently floating above the Sun's photosphere. They are conclusively considered to owe their origin to the Sun; but, from their comparative faintness, they had been only observed when the solar disk was totally obscured by the Moon, until after the eclipse of August, 1868, when the spectroscopic observations of M. Janssen, at Guntoor, in India, and of Mr. Lockyer, in England, revealed them in full sunlight.

Photographs were taken of the different phases of the eclipses of 1860 and 1868, in the former year by Mr. De La Rue, and in the latter by Major Tennant, at Guntoor, and by a German party at Aden. In Mr. De La Rue's photographs the progress of the Moon over several prominences is distinctly marked. This direct evidence of the movement of the Moon over the prominences is much stronger than that of the eye-observations of the most experienced observers; for, during the totality, one's mind is so engrossed with so many unexpected novelties as to make it impossible to grasp every detail in the short interval of time occupied by the total obscuration.

Attention was first directed to these prominences by Vassenius, who observed them in Sweden during a total eclipse which occurred in 1733. No particular attention was, however, paid to them till the eclipse of July 8th, 1842, when they were again noticed by Mr. Baily at Pavia, and by Mr. Airy at the Superga hill, near Turin. In this eclipse Mr. Baily saw three large protuberances, having the appearance of mountains of prodigious elevation, of a red colour, tinged with lilac or purple. They somewhat resembled the sunny tops of the Alpine mountains when coloured by the rising or setting Sun. Mr. Airy also observed three red prominences, and, what is most remarkable, they were seen by Mrs. Airy without the aid of a telescope.

These curious protuberances on the Moon's limb caused such universal interest to observing astronomers that, on the occurrence of the eclipse of July 28th, 1851, a large number visited Norway, Sweden, and Eastern Prussia, over which countries the narrow dark band of totality passed. The Government sent out a special expedition, consisting of the Astronomer Royal, Mr. Humphreys, and the author. The Astronomer Royal observed the eclipse at Göttenburg, Mr. Humphreys at Christianstadt, and the author at Christiania. Of the private observers, Mr. Hind and Mr. Dawes were stationed at Rævelsberg, near Engelholm, Sweden; Mr. Adams at Frederiksværn, Norway; Mr. Lassell at Trollhätten Falls, near Göttenburg; and Mr. Carrington at Lilla Edet, on the Göta river, Sweden. Other observers were scattered over the narrow zone of totality from the western coast of Norway to Eastern Prussia.

Nearly every astronomer made successful observations of the different phenomena. One of the prominences was of such a peculiar form as to attract the attention of every observer. This curiously-formed prominence was of a horned shape, curved in the direction of the lower portion of the Moon. Its hooked appearance made such an impression on the author's mind that he formed a fair estimate of its magnitude. He was able easily to do this, because there were some parallel wires inserted in the telescope, the value of the interval between which he had accurately determined a day or two previously. The height of this hooked prominence was found to be about 45,000 miles, and its breadth at the base 15,000 miles. Its colour was rose-pink, but of a very deep shade. On the appearance of the first direct ray of light from the solar disk, this protuberance, together with two others of smaller dimensions, suddenly disappeared, no trace of them being visible after the formation of the beads at the end of the totality. To the Astronomer Royal the large prominence appeared like a bomerang, the colour being full lake-red in some parts, the remaining portion being nearly white. This prominence was seen by the Astronomer Royal, and by some of his friends, with the naked eye. He also noticed many other prominences, and a red detached cloud, in the form of a balloon, floating above the Sun's photosphere. Mr. Hind and Mr. Dawes also saw a detached cloud, which is identified with that seen by Mr. Airy. Referring to the large prominence which was universally observed from Western Norway to Eastern Prussia, Mr. Dawes remarks, " In shape it somewhat resembled a Turkish cimeter; the northern edge being convex, and the southern concave. Towards the apex it bent suddenly to the south. Its northern edge was well defined, and of a deeper colour than the rest, especially towards its base. I should call it a rich carmine. To my great astonishment, this marvellous object continued visible for about five seconds after the Sun began to reappear. It then rapidly faded away, but it did not vanish instantaneously. From its extraordinary size, curious form, deep colour, and vivid brightness, this protuberance absorbed much of my attention."

In the eclipse of the Sun of July 18th, 1860, to observe which the Admiralty granted the free use of H.M.S. "Himalaya" for the conveyance of the astronomers and their instruments to the north coast of Spain, the red prominences were successfully observed. It is unnecessary to describe in detail the accounts of the forty-three *savans* who were included in the "Himalaya" party; nor even of the many eminent foreign astronomers who visited Spain for the same purpose. A single example will be sufficient. We shall, therefore, merely give the substance of an extract from the remarks of the well-known discoverer of minor planets, M. Goldschmidt, who was stationed at Vittoria. He states in his account that the most imposing as well as complicated of all the prominences, called by him the chandelier, was grand beyond description. It rose up from the Moon's dark limb, appearing like slender tongues of fire, and of a rose colour; its edges purple and transparent, through which the interior of the prominence could be distinguished: in fact, it could be distinctly seen to be hollow in structure. Shortly before the end of the totality a slight display of light in the shape of a fan was seen to be emitted from the summits of these rose-coloured and transparent sheaves of light, giving to the prominence the *real* appearance of a chandelier. Its base, which at the commencement of the totality was noticed to be very decidedly on the black limb of the Moon, became slightly less so, when the whole looked more ethereal or vapourish. The jets of light which proceeded from the summits of the prominence disappeared with the shooting forth of the first ray of the Sun; but it was not so with the protuberance itself. An instant before the end of the totality several small prominences were seen to be lying close to each other on the right of its base, forming a square. Two others of the same size were also seen on the left side of the base, as viewed by a telescope which inverted.

In the total eclipse of the Sun observed in India in August, 1868, one very remarkable protuberance, as in 1851, was universally seen. It was remarked by Dr. Weiss, who was stationed at Aden, that it was in length equal to the eighth part of the Sun's radius, or fifty-three thousand miles. At Guntoor, Major Tennant recorded the height as 3′ 20″, or upwards of eighty thousand miles. This great prominence was the one selected for the analysis of its spectrum by M. Janssen, Major Tennant, Lieut. Herschel, and M. Rayet, at different stations on the central line of shadow; the result, announcing the self-luminous gaseous nature of its constitution, was telegraphed to Europe on the same day.

With reference to the material of which these prominences are composed, M. Le Verrier had suggested that the Sun may be simply a luminous body on account of its high temperature, and that it is probably covered by a continuous bed of rose-coloured matter, the existence of which has been so abundantly proved from the observations made since 1842. This suggestion of M. Le Verrier has been completely verified by the spectroscopic discoveries of Mr. Lockyer and M. Janssen. Mr. R. Grant and M. Littrow had also expressed their opinion that the observations of 1842 and 1851 clearly indicated the existence of a narrow layer of luminous matter exterior to the ordinary photosphere of the Sun.

Though the appellation of "total darkness" has been given to the intense gloom which always takes place when the Sun is totally hid, yet no modern observer has found insufficient light to prevent large print being read at the ordinary distance from the eye. But the unnatural gloom, in which everything is wrapped, affects in some degree all who witness it. In 1842 Mr. Airy remarked that the appearances were very awful. The gloom increased every moment. The candle seemed to blaze with unnatural brilliancy. A large cloud in the zenith, whose appearance he had not particularly remarked, but which was probably of a cumulostratus character, became converted into a black nimbus, blacker, if possible, than pitch, and seemed to be descending rapidly; its aspect became terribly menacing, and seemed to be animated. Mr. Airy says, "Of all the appearances of the eclipse, there is none which has dwelt more powerfully upon my imagination than the sight of that terrible cloud." The same astronomer has stated that in 1851 the darkness was even more striking than in 1842. At Christiania, in 1851, the darkness was not similar to that of night, for the outlines of mountains at least fifteen miles distant were faintly visible, and the title-page of a book could be read ten or twelve inches from the eye, though there was a difficulty in doing so. But the general appearance of the landscape was so extraordinary that, at this lapse of time, it is firmly fixed on our memory. Any of our readers who have journeyed to the upper part of the Christiania Fiord must have noticed the numerous small islands which are scattered about, particularly near the city. The effect of the darkness on this mixture of land and water was very peculiar; the colour of the water being a deep purple, and the islands a dusky yellow. During the progress of the eclipse the sky was generally covered with cloud of different densities, giving much anxiety to the observers. The Sun was, however, comparatively free from cloud during the totality. The awful appearance of the terribly black cloudy portions of the sky, and the intensely gloomy look of the remaining portion, cannot easily be forgotten.

The following account of the darkness, as viewed by a lady at Christiania, gives the impressions of a non-scientific observer. Though somewhat imaginative, still its general truthfulness may be relied upon:—

"Our very hearts stood still; Nature herself grew suddenly silent; the songs of birds ceased; the animals huddled together, and cowered in silence; the darkness swept on, swept over us, wrapped its wings around us; a strange greenish yellow hue mingled with it, and gave it the most supernatural aspect. The horizon wore a belt of that greenish yellow hue; the vegetation around us assumed it; the human faces in which I looked reflected it. The Fiord, with its waters and rocky islets, was covered in that strange pall; and through the mysterious and impressive gloom uprose the tall pines from these islets, looking like gigantic spectres rising from out of chaos—a paler, yellower shade than the darkness around them. All was unearthly-looking, but unspeakably grand, full of awe and solemnity. How long that darkness lasted I know not; mine were not scientific observations; but, quickly as it had travelled, it moved not quicker than the thoughts and impressions of the human mind. The darkness had not gathered round us as other darkness does: it had swept on from one quarter of the heavens to the other; and we saw it coming from one side, while the other was still light. Now it seemed to rise up at once from us, as if it lifted its great wings and gathered itself up. We saw from whence it came; we saw not whither it went."

It was remarked by Captain (now Colonel) Biddulph, R.A., who observed, in 1851, at Dröbak, Norway, that during the time the Sun was covered the light reflected from most of the common objects near him was of a deep blue purple. He had collected a few wild flowers, blue and yellow, and some red strawberries, which he formed into a group. He found that the red was not reflected at all, blue tolerably, yellow not. Some blossoms of clover, which were more of a crimson purple, were slightly distinguishable. Captain Biddulph considered that the darkness was greater than at the preceding midnight. The darkness of the midnight of southern Norway in the month of July is not, however, very intense.

Mr. Hind, at Rævelsberg, noticed that, five minutes before totality, the whole landscape appeared gloomy and unnatural; the sky being intensely blue and the air strikingly chilly. After a short interval, the southern heavens became of a deep sombre purple, or purplish grey, the corona distinct to the naked eye; while in the zenith and northwards, the sky appeared as if it had closed in, and was of an intense violet colour. Near the horizon, from the N.N.W. to E.N.E., the sky was occupied with bands of a deep orange red, with intermediate spaces of a purple hue, forming the most astonishing and appalling phenomenon that it is possible to imagine.

In the eclipse of 1860, Mr. Airy considered that the darkness was not so great as on the two previous occasions. Mr. De La Rue, who observed this eclipse at Rivabellosa, near Miranda de Ebro, observes that, when he had once turned his eyes on the Moon, encircled by the glorious corona, then on the novel and grand spectacle presented by the surrounding landscape, and had taken a hurried look at the wonderful appearance of the heavens, so unlike anything he had ever before witnessed, he felt so completely enthralled by the scene that he had to exercise the utmost self-control to tear himself away from a view at once so impressive and magnificent. It was with a feeling of regret that he was obliged to give up the sight and continue his self-imposed duties.

Mr. Airy, with several members of his family, observed this eclipse at Hereña, near Miranda. At this station Mercury, Venus, and Jupiter, and six of the principal stars, were distinctly seen. What struck Mr. Airy most, concerning the darkness, was the great brilliancy of Jupiter and Procyon so near to the Sun. It was impossible that they could have been seen at all, except under the circumstance of total absence of illumination in that part of the atmosphere through which their light passed. At the moment of totality, according to Mrs. Airy, the whole air was at once filled with darkness, which had been seen rapidly approaching from the north-west. A sensation of cold was felt, and Mrs. Airy was glad to wrap herself up in a Scotch plaid. At the end of totality the dark shadow was seen distinctly sweeping along the valley to the south-east, a path of darkness, and the clear daylight breaking out behind it:

De Louville states that the darkness in London, in 1715, was so great that three planets, and several stars down to the second magnitude, were visible. He has also recorded that the London atmosphere of 150 years ago had, even then, an unenviable notoriety for its impurity; for he says that "it was a piece of good fortune at this time to have found in London an interval of clear sky. It is, indeed, so rare to see the Sun in that city, either from cloud or smoke, that, during the month in which I resided there, I do not believe the sky was clear on more than three days."

During the eclipse of 1868, the observers report generally that the darkness was much less than what was expected, considering the great duration of the total obscuration. Major Tennant remarks that at Guntoor the colour of the sky was not half so gloomy as at Delhi during the partial eclipse of 1857.

Nearly all the American astronomers who observed the total eclipse of August 7th, 1869, under the most favourable circumstances possible, have remarked that the darkness during the total obscuration of the Sun was considerable, in appearance very similar to that of a moonlight night. At Springfield, Illinois, where the sky was perfectly cloudless, several of the planets and fixed stars were distinctly visible, including Mercury, Venus, Mars, Saturn, Arcturus, and Regulus. A special, but unsuccessful, search was also made for the supposed intra-Mercurial planet, Vulcan.

No favourable total eclipse of the Sun has occurred, without many anecdotes being related concerning the effects produced by the sudden darkness on the peasantry, animals, and plants. The following is a translation of one which appeared in the "Journal of the Lower Alps" of July 9th, 1842:—A poor child of the commune of Sièyes was watching her flock when the eclipse commenced. Entirely ignorant of the event which was approaching, she saw with anxiety the Sun darken by degrees; for there was no cloud or vapour visible which might account for the phenomenon. When the light disappeared all at once, the poor child, in the height of her terror, began to weep and call out for help. Her tears were still flowing when the Sun again sent forth his first ray. Reassured by the aspect, the child crossed her hands, exclaiming, in the *patois* of the province, "O beou souleou!" (O beautiful Sun!) Mr. Fox Talbot, who, in 1851, was at Marienburg, Prussia, has related the following:—On the morning of the eclipse, a Prussian officer, with whom he had been conversing, remarked that the only observation he intended to make would be this—that he should mount his horse, and ride alone by the side of the river, to see what effect, if any, would be experienced by his horse. On Mr. Talbot meeting the officer in the evening of the same day, the latter was asked whether he had put his purpose into execution. He replied that he had done so, and had very nearly lost his life in consequence; for the animal was seized with such a panic terror on the extinction of the Sun that he was scarcely able to master him, and both steed and rider were in the utmost danger of being precipitated into the river.

The following illustration of superstitious fear, which frequently takes hold of the minds of the peasantry, came under the notice of the author in 1851, at Christiania. A respectable old woman was in the habit of daily supplying the principal hotel with wild strawberries. On the Saturday preceding the eclipse, she begged of the landlord to take an extra quantity of strawberries, as she had been told that the philosophers at the college had said that on Monday the Sun would disappear from the heavens. Thinking that this was a sign that the end of the world was approaching, she could not think of coming into the city on that day, for she wished to remain at home to say her prayers. In contrast to this, Lieutenant Krag saw an old woman quietly light her candle, and with perfect indifference continue her work.

Of animals, the greatest consternation is generally seen in birds. This has been noticed in every eclipse. In 1715 De Louville remarked that shortly before the Sun was totally eclipsed the cocks began to crow as at daybreak; during the darkness they were silent; but as soon as the Sun reappeared they recommenced with increased animation. At the total obscuration, birds hid themselves in extraordinary places, but seemed themselves again on the return of the light. Fowls were seen to prepare for roosting, as in the shades of evening. At Christiania, in 1851, a bird in a neighbouring bush was keeping us company with a merry song. At the instant of totality our friend suddenly ceased; but when the first solar ray again illumined the heavens he recommenced his former song. In the midst of the darkness some crows rose together and flew irregularly about, uttering what appeared to us at the time unearthly cries.

An amusing anecdote on the effect produced on an old negro and his flock of hens during the total eclipse of the Sun of August 7th, 1869, has been related by Dr. C. H. F. Peters, director of the observatory of Hamilton College, Clinton, New York. The negro was requested to pay particular attention to the movements of the hens, for he was told that at a quarter to five they would all go to roost. After the totality was over, he came to Dr. Peters, evidently under great excitement. "How was it?" said the doctor. "Beats de debbil," said the negro. "When de darkness come, ebry chick'n run for de hole in de barn. De fust ones got in, and de next ones run ober one anudder, and de last ones dey just squat right down in de grass. How long you know dis ting was a coming?" "Oh! I reckon we knew it more than a year," said the doctor. "Beats de debbil! Here you away in New York knowed a year ago what my chick'ns was gwine to do dis bery afternoon, an' you nebber see de chick'ns afore nudder!"

The effect of the darkness on plants and flowers has been observed frequently. The leaves and petals of most sensitive flowers have been seen to close, particularly the flowers of the silk-acacia tree, convolvulus, and other plants of that nature.

THE MOON.

"By thy command the Moon, as daylight fades,
Lifts her broad circle in the deepening shades;
Arrayed in glory, and enthroned in light,
She breaks the solemn terrors of the night;
Sweetly inconstant in her varying flame,
She changes still, another, yet the same!
Now in decrease, by slow degrees she shrouds
Her fading lustre in a veil of clouds;
Now of increase, her gathering beams display
A blaze of light, and give a paler day.
Ten thousand stars adorn her glittering train,
Fall when she falls, and rise with her again;
And o'er the deserts of the sky unfold
Their burning spangles of sidereal gold;
Through the wide heavens she moves serenely bright,
Queen of the gay attendants of the night."

BROOME.

EXT to the "greater light that rules the day," the most conspicuous object in the heavens, as well as the most attractive, is the constant attendant of the Earth in its annual course around the Sun, the Moon, or the "lesser light that rules the night." It is probable that many people even look upon our nearest companion in space with far more interest than upon the brilliant Sun itself; not so much, however, for the delineations of light and shade exhibited on the lunar disk, as for the continual change of form to which it is subject on account of its rapid movement around the Earth, from which it is distant about 238,000 miles.

The time occupied by the Moon in performing a complete journey around its primary is 27^d 7^h 43^m $11·461^s$, called its sidereal period. The lunar month is longer than the sidereal period by 2^d 5^h 0^m $51·41^s$, in consequence of the progressive motion of the Earth in its orbit in the interval between two consecutive conjunctions of the Moon. Our satellite must therefore pass through an additional arc, equivalent to that apparently traversed by the Sun since the previous conjunction, before it can assume the same phase, or be again in a line with the Sun and Earth. The interval of a lunar month is known as the synodical period.

Astronomers of all nations have made the Moon an object of observation and research to a far greater extent than any other member of the solar system. Some have made careful delineations of the lunar disk, tracing with the utmost accuracy the form and relative distribution of the numerous features seen on the surface. Others, again, have investigated the peculiar motion of the Moon in its orbit, by means of the accurate lunar observations made at fixed observatories, principally, however, at

TELESCOPIC FACE OF THE FULL MOON.

the Royal Observatory, Greenwich. These researches, usually undertaken by the highest class of mathematicians, are of the greatest importance to the master-mariner, who, by the use of the positions of the Moon given in the Nautical Almanack, and computed from tables formed directly from the labours of the astronomer, is enabled to guide his ship without fear, by his sextant-observations, from one side of the globe to the other. For from a comparison of the observed angular distances of the Moon from the principal stars and planets, with the corresponding calculated lunar distances given in the Nautical Almanack, the chief officer knows exactly in what longitude his ship is situated, even when land has not been perceived for several weeks. The observations of

TAKING A LUNAR DISTANCE.

the Moon for the rectifying of the lunar tables have been, and continue to be, the principal astronomical work for which the Royal Observatory has been so justly celebrated from the days of Flamsteed to the present time.

Our knowledge of the lunar surface is confined to one hemisphere—an effect caused by the coincidence of the time of rotation of the Moon on its axis being exactly equal to the time occupied in its revolution around the Earth; the same face is consequently always turned towards us. The Earth is therefore invisible from one half of the lunar surface, while from the opposite hemisphere it appears constantly in the firmament, by day and by night, and always in the same position, changing only by an alteration in the station from which it is viewed. When illuminated by the Sun, the Earth has the same appearance from the Moon as the Moon has to us, but of greater magnitude in the proportion of seventy-nine to twenty-one. It would go through the same phases, and complete the series of them in a similar manner by which the lunar phases are regulated; but when the Moon is full to the Earth, the Earth is new to the Moon, and *vice versâ*, and when the Moon is a crescent, the Earth is gibbous, and *vice versâ*, the one being always the complement of the other.

Most persons, even in their youthful days, have watched with interest the gradual change from day to day in the form of the Moon, from the narrow and

THE LUNAR CRESCENT.

beautiful crescent of a few days old to the full round disk at opposition. There are many, however, who see these continual monthly changes without knowing, or caring to inquire, the causes which produce them. While the Moon is performing its revolution round the Earth its illuminated hemisphere will of necessity be always turned towards the Sun. When the Moon is therefore viewed from the surface of the Earth, the lunar phases must depend solely on the relative positions of the three objects, and as these relative positions change, so the amount of the illuminated disk of the Moon will vary likewise. For example, at conjunction, when the new Moon is situated between the Earth and the Sun, the solar rays fall entirely on that portion of the Moon which is invisible to us, consequently the hemisphere which is turned towards the Earth has no illumination from the Sun, and therefore it cannot be perceived by us. But, a few days after conjunction, in consequence of the Moon's progress in its orbit and of its corresponding rotation on its axis, our satellite will appear, soon after sunset, as a narrow crescent near the western horizon. At this time the greater portion of the illuminated hemisphere is still turned away from us, but a small part of the enlightened disk has come into view. As the Moon pursues its course, the illuminated portion of the disk is gradually turned more and more towards the Earth till it appears as a half-moon, when one quarter of a revolution has been completed. This phase is known popularly as the "first quarter." In this position the hemisphere directed to the Earth is one half illuminated, and one half unilluminated. Between this and opposition the form of the visible Moon is gibbous. At full Moon the enlightened disk is opposite the dark hemisphere of the Earth, and is now always in its most perfect form, passing the south meridian at its greatest altitude at midnight. In the second half of the lunation, the enlightened portion of the Moon diminishes in the same ratio as it increased in the first half, the left-hand side of the disk being now visible, and the direction of the horns reversed. When the illuminated disk is reduced again to the form of a half-moon, the Moon is stated in all almanacks to be in its "last quarter." The diminution of brightness gradually continues until we can see only the crescent Moon near the eastern horizon. In a few days more, the three objects are again in nearly a direct line, the Moon being the intermediate body, and, as we have before stated, the illuminated disk being completely turned away from us at this time, our satellite is necessarily absent from our evening and night skies.

It must be borne in mind that throughout these apparent changes, the same actual face is always directed to the Earth, whether it be illuminated or non-illuminated.

If the surface of the Moon be submitted to a careful examination, even without the assistance of a telescope, distinct and definite lineaments of light and shadow are plainly visible upon it. These appearances never seem to change; their relative positions never undergo any variation; in fact, the features now exhibited are the same as those described in the earliest records. We may naturally ask the question, What is the character of these lineaments of light and shade visible to the unassisted eye? For an answer we must have recourse to the telescope. By means of telescopic aid the observer is enabled, with a marvellous precision, to resolve these lunar characteristics into definite rocks, mountains, volcanoes, and plains. They are scattered over the visible surface in every direction, being of variable form and magnitude. The mountains, or craters, are generally known and identified by the name of some ancient or modern astronomer, including Eratosthenes, Ptolemy, Tycho, Copernicus, Kepler, Huygens, Newton, Flamsteed, Bradley, Airy, Herschel. Several parts of the lunar surface, of considerable extent, are mostly free from mountains, though not entirely; these apparent plains were formerly supposed to be seas or oceans. They are distinguished by such names as the Mare Nubium, the Mare Crisium, the Mare Humorum, the Oceanus Procellarum, etc., their number being thirteen.

For the purpose of illustration, it will suffice to describe the peculiarities of a few only of the principal mountains, which will give some idea of the magnitude of these lunar phenomena. There is much in common in their general form, but they are of every conceivable size. We will begin with Tycho (seen in the lower part of the figure on page 241), which is always considered one of the most interesting of these objects.

The inclosed area of Tycho is about fifty miles in diameter, being nearly circular. The height of the western ridge of mountains surrounding the area is about 17,000 feet, and that of the eastern ridge about 16,000 feet. Within the inclosure, a central mount, about 4,500 feet high, and a few lesser hills, are situated. The district in its immediate neighbourhood is exceedingly rugged; craters, peaks, and long ridges of mountains may be observed in every direction for a considerable distance. From it flows a series of luminous streaks or rays, which extend over a large portion of the Moon's disk, commencing about twenty miles from the exterior of the circular ridge of the crater. These streaks are not visible until the Sun's rays fall upon the

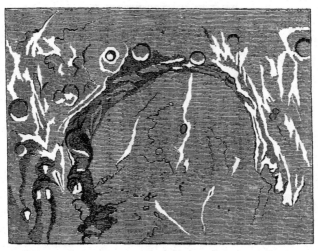

PORTION OF THE LUNAR DISK.

region of Tycho at an angle of twenty-five degrees or less; therefore, the more perpendicularly the rays fall upon it, the more visible the streaks will appear. Consequently they are seen to the best advantage at full Moon. Astronomers have many conjectures as to the nature of these luminous streaks; by some they have been considered to be mountains, by others it has been thought that they were, at some period, streams of lava, having flowed originally from the great central crater. It has also been supposed that, by a sudden volcanic upheaving of the lunar crust, the effect has been produced, similar to that of a pane of glass, or a sheet of ice, broken by a pointed hammer. The part immediately above that, where the upheaving agency is situated, would probably be the point where the greatest disruption would take place, from whence the cracks would radiate, allowing the lava to flow freely in gentle streams, according to the size of the opening, throughout its whole length. Mr. Hind remarks that "the mere fact of their diverging from the great crater, Tycho, proves that it was the focus of the volcanic outbreak, whenever it may have occurred." These are speculations, however; and there is in reality too little known to warrant any definite conclusion.

Copernicus is another large annular mountain of the same class as Tycho, though of somewhat larger dimensions, being about fifty-five miles in diameter, the highest part of the external ridge of mountains being about 11,000 feet. This crater is most easily seen when the Moon is in its first quarter, for at that time the shadows produced by the Sun's rays, being intercepted by the ridge of mountains on the western side of the crater, are projected on the inclosed area, while the shadow of the eastern ridge darkens, for a considerable distance, the exterior plain near the mountain. These shadows are generally very well defined and extremely black. Radiating streaks flow from Copernicus, in a similar manner to those belonging to the region of Tycho, but not to the same extent.

N.W. BOUNDARY OF THE MARE SERENITATIS.

The remaining lunar mountains around which this extraordinary radiation is manifested, are Kepler, Proclus, Olbers, Aristarchus, and Anaxagoras. Similar phenomena, but in a manner less distinctly marked, are visible around a few others.

Eratosthenes is a remarkable mountain, of the annular class, situated near the extremity of a long range of immense protuberances, known as the Apennines. The crater of this beautiful specimen is thirty-seven miles in diameter, a precipitous rock, nearly 16,000 feet in altitude, being placed in the centre. When the Moon is near the first quarter, Eratosthenes presents a very beautiful appearance, if viewed with a telescope furnished with a moderately magnifying power. The shadows projected on the plain by the exterior ridges of the mountains which form the crater, are very distinctly defined, and afford a most interesting subject for the amateur astronomer.

Longomontanus is another celebrated circular range of mountains, being eighty miles in diameter. The eastern and western ridges rise to the height of 12,000 or 13,000 feet above the level of the inclosed plain. The numerous craters of small magnitude which lie in close proximity to this mountain, are sometimes concealed by its shadow, which is sufficiently large to cover all near objects. The surrounding region is savage and rugged in the highest degree; distinctly proving that these extinct volcanic remains must have resulted from a long succession of convulsions.

Gassendi, as seen with a powerful telescope, is a favourite object with astronomers. This remarkable series of craters consists of two stupendous chains of mountains; the outer ridge, which varies from 3,500 to 5,000 feet in height, being about 60 miles in diameter. The area of the inclosure of this extensive range is 2,800 miles, in the centre of which a curious mountain furnished with eight peaks is situated, while numerous others of less elevation are scattered about the inclosure. The inner range of mountains is 16 miles in diameter.

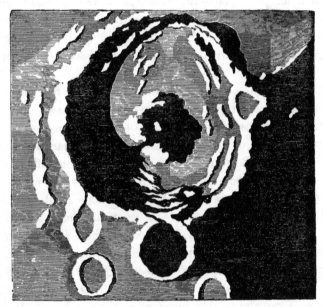

THE LUNAR MOUNTAIN GASSENDI.

The altitude of the lunar mountains is generally considerable; but the greater number do not present any remarkable appearance. About twenty exceed the height of 16,000 feet; one of the highest, Newton, being about 24,000 feet. This latter may be compared to some of the loftiest summits of our South American or Himalayan ranges.

Some of the objects which we have described can be detected by a sharp eye; but, when viewed by a powerful telescope, they unfold to the mind scenes of marvellous beauty. We have seen them through the magnificent telescopes of the Great Equatorial, at Greenwich, and the Northumberland Equatorial, at Cambridge, using a high magnifying power. The ridges inclosing the plains of the ringed mountains appear projected towards the eye, as in a stereoscopic view. Rocks upon rocks are piled on each other in different layers; the shadows of the mountains are thrown upon the broad plains with an intense blackness; whilst the rugged nature of the whole lunar surface has a most striking effect. To all appearances, it seems composed of one great solitary rocky waste, unfit for the habitation of living creatures.

We have already alluded to the large plains, on which the volcanic character is much less evident, though some of the principal mountains and many smaller ones are situated within their boundaries. These plains have sometimes been called seas, as their names imply; but this nomenclature does not exactly state the true nature of these apparently flat portions of the Moon's surface. The dimensions of a few of these plains we will describe very briefly. The Mare Imbrium is about 680 miles from north to south, and 750 miles from east to west. The Mare Crisium is 280 miles in diameter in one direction, and about 350 miles in the opposite. The Oceanus Procellarum is the largest of the lunar plains, covering a surface of 90,000 square geographical miles. The Mare Serenitatis is an elliptical spot, with an average diameter of about 430 miles. The remaining nine are all of considerable extent. Most of the lunar phenomena are seen to perfection with an ordinarily good defining telescope fitted with an eye-piece of low power.

The details of many of the principal objects on the lunar face are very clearly shown on some of the beautiful photographs which have been taken within the last few years. Among the most successful, those of Mr. Warren De La Rue may be specially mentioned.

Among the many proofs of the non-existence of a lunar atmosphere, it may be mentioned that no water can be seen; at least, there is not a sufficient quantity in any one spot so as to be visible from the Earth. Again, there are no clouds; for if there were, we should immediately discover them by the variable light and shade which they would produce. But one great proof of the absence of any large amount of vapour being suspended over the lunar surface is the sudden extinction of a star when occulted by the Moon. The author has been a constant observer of these phenomena, and, though his experience is of long standing, he has never observed an occultation of a star or planet, especially at the unilluminated edge of a *young* Moon, without having his convictions confirmed that there is no appreciable lunar atmosphere. In occultations of this kind the star is seen to approach the dark edge the Moon, which is very visible, owing to the reflected light of the Earth shining on it. If there be any atmosphere of reasonable density around the Moon, small stars would disappear, or at any rate, they would considerably diminish in their intrinsic brightness before they reach the unilluminated edge. But this is never the case: for they are seen in their full lustre till their sudden extinction at the Moon's limb. Professor Challis has subjected the results of a large number of these observations to a severe mathematical test; but he has not been able to discover the slightest trace of any effect produced by a lunar atmosphere.

The mass of the Moon is about the $\frac{1}{81}$st part of that of the Earth, or, it would require eighty-one Moons to make a globe of corresponding weight to that of the world in which we live.

248

THE TERRESTRIAL PLANETS.

"With what an awful world-revolving power
  Were first the unwieldly planets launched along
  The illimitable void! There to remain
  Amidst the flux of many thousand years,
  That oft has swept the toiling race of men
  And all their laboured monuments away,
  Firm, unremitting, matchless in their course;
  To the kind-tempered change of night and day,
  And of the seasons, ever stealing round,
  Minutely faithful. Such the all-perfect Hand,
  That poised, impels, and rules the steady whole."
                                  THOMSON.

AKING the planets together, they naturally fall into three groups, the Terrestrial Planets, the Minor Planets, and the Major Planets. The immense space which separates the first group from the last is sufficient to make a visible distinction between their respective planets. Mercury, Venus, the Earth, and Mars, have very much in their physical features common to each other, and many remarkable analogies are also exhibited in their movements. In like manner, the greater magnitude of the four distant planets, each the centre of a complex system of its own, presents an aspect so different from the small, but comparatively quickly-moving, planets nearer the Sun, that it requires but little consideration to include them in one class.

Before giving a few popular and familiar notes on each of the three classes of planets, we must refer to an object which was originally believed to be a new one by some of the principal astronomers, including M. Le Verrier, the distinguished Director of the Imperial Observatory of Paris, who undertook, at the time of the supposed discovery, to examine all the circumstances of the observation on the spot. Without going into any detail, it may be stated that a small dark body was seen to pass across the solar disk on March 26, 1859, by a M. Lescarbault, a physician at Orgères, in the department Eure-et-Loire, France. This dark body was totally different from an ordinary solar spot, being perfectly round, with a planetary appearance. M. Lescarbault, expecting, or at least hoping, that he might see the phenomenon again, kept his observation as a secret for months. It was not until the end of September that M. Le Verrier was first made acquainted with the alleged observation. The subject was of intense interest to him, for he had previously announced that, to make the theoretical calculations of the positions of Mercury agree with the observations, corrections must be applied for the perturbations produced by an unknown planet between Mercury and the Sun. Although M. Le Verrier found the village doctor neither a professed astronomer nor mathematician, and that his notes were recorded in the rudest manner possible, yet he came to the conclusion that the dark body observed by M. Lescarbault was not a hypothetical object, as might be supposed, but that it was really a true planet; that its distance from the Sun was about fourteen millions of miles; that its revolution round the Sun occupied twenty days; and that its orbit was inclined to the ecliptic twelve or thirteen degrees. Since the alleged discovery, the suspected planet has been systematically looked for at several of the principal observatories, at the times when the orbit of the object ought to have been projected on the Sun's disk. No trace of it has been seen. M. Liais even goes so far as to say that he was engaged in the observation of solar phenomena at Rio Janeiro at the identical moment of M. Lescarbault's supposed discovery, and that he is perfectly certain that no object of a planetary nature was passing over the Sun at the time. Meanwhile this apocryphal planet has received the name of Vulcan.

MERCURY.

The planet Mercury was known to the ancients, although it is only occasionally visible to the naked eye without some optical aid. There are times, however, when it is even a conspicuous object, but from its low elevation it is very likely to pass unnoticed. When the planet is in that part of its orbit known as the greatest elongation, or at the greatest apparent distance from the Sun as viewed from the Earth, it shines with a rosy lustre equal to that of any star of the second magnitude, notwithstanding the strong twilight in which it is always enveloped. When it is in its extreme eastern elongation, it is visible in the evening near the western horizon; but when the planet is on the opposite side of the Sun, it can be seen only in the mornings near the eastern horizon. The discovery of Mercury by the ancients is a proof of the steady and careful observation of the heavens by the astronomers of the classic ages. There have been many astronomical observers of the last and present centuries, who have never been able to detect this planet with the unassisted eye; and it is said that Copernicus, the great expounder of the modern system of the Universe, was never able to obtain a view, although he made several attempts to do so. Gassendi, however, attributes the failure to the mist and vapour so very prevalent along the banks of the Vistula. Tycho Brahé, from the island of Huen in the Sound, frequently records in his journals the observation of Mercury with the naked eye.

Mercury revolves around the Sun in about eighty-eight days, at an average distance of 35,550,000 miles. In consequence of its orbit being a sensible ellipse, its greatest and least distances differ nearly fifteen millions of miles. Its diameter is about 3,000 miles, and its volume or bulk the nineteenth part of that of the Earth. It is supposed to rotate on its own axis in 24^h 5^m 28^s, but on account of the proximity of the planet to the Sun, no spot or marking has been satisfactorily seen which might be used for a determination of its rotation. Schröter, in the beginning of this century, fancied that he had evidence of the existence of mountains on the surface of Mercury, from which the above value was obtained. Sir John Herschel, however, considers that our knowledge of the physical condition of this intra-terrestrial planet consists of little more than that it is globular in form, exhibiting phases similar to Venus and our own Moon.

When Mercury is in superior conjunction, that is, when it is beyond the Sun relatively to the Earth, the three bodies being in a line of which the Sun is the central one, Mercury and the Sun will be on the meridian at the same time. When it is in inferior conjunction, these two bodies will also pass the meridian at the same instant, but the planet is then between the Sun and the Earth. It is evident, therefore, in the latter instance, that if a line be drawn from the Earth to the Sun, passing through Mercury, this planet must be seen from the Earth projected on the solar disk, and what is called a "transit of Mercury" must take place. Now, this phenomenon has been frequently observed, but not until the year 1631, in which year Gassendi was the first person who saw Mercury projected on the Sun. He employed for his purpose a sheet of white paper, placed in a chamber from which all stray light was excluded. The sight of the black disk of the planet filled the zealous astronomer with the greatest enthusiasm. "I have seen," he says, "that for which alchymists have sought with so much ardour, I have observed Mercury in the Sun." Shakerley observed the next transit in 1651. After the invention of the telescope in the beginning of the seventeenth century, the phenomenon became an easy subject of observation. During the present century, the transits which occurred in the years 1802, 1832, 1845, 1848, 1861, and 1868, were all favourably seen.

When Mercury is on the solar disk, numerous observers have remarked that the dark body was surrounded by a nebulous ring, probably from the effects of contrast, others have noticed a small luminous spot on the planet. Many, however, with larger telescopes, have been unable to perceive anything peculiar either on or around the black disk. The most valuable observations of transits of the inferior planets over the Sun are the careful records of the exact instants of time when the planet enters on and recedes from the solar disk, usually termed the ingress and egress. The most important moments to note, however, are the exact times of the internal contacts, or just when a small thread of light is seen between the edges of the dark planet and the Sun. At such times a very curious phenomenon has been observed, both in transits of Venus and Mercury. At the ingress, as the planet advances over the solar disk, a few seconds after the apparent internal contact, a dark line, called the "black-drop," has been seen to shoot out from it, or the edge of the planet nearest the solar edge becomes elongated or pear-shaped. At the egress it is observed before the apparent internal contact. This phenomenon was first noticed in the transit of Venus of 1761, and again in 1769. In transits of Mercury it has also been frequently observed, but in the latter the effect is less in proportion to the difference in the magnitudes of Venus and Mercury. During the last transit of Mercury on the morning of November 5th, 1868, the formation of the black-drop preceding the apparent internal contact of the edges of the Sun and planet at the egress, was very clearly observed at the Royal Observatory, Greenwich. In the telescope of the great equatorial, the phenomenon was first seen as a thin dark filament stretching across the luminous line still existing between the limbs of the two bodies. In the altazimuth telescope, the existence of the black-drop was indicated by the planet assuming an elongated or pear-shape, tapering almost to a point at the Sun's edge. In smaller telescopes, the same phenomenon, under different phases, was exhibited. Its cause may be explained by supposing that the great irradiation, or spreading out, of the intense light of the Sun, cuts off or hides from view, during the transit, a portion of the planetary disk; that part which is hidden becoming suddenly visible as a dark line as soon as the planet reaches the true edge of the Sun unaffected by irradiation, which occurs when to the eye a luminous thread seems still to separate the two edges. The remaining transits of Mercury during this century will take place on May 6th, 1878; November 7th, 1881; May 9th, 1891; and November 10th, 1894.

VENUS.

When the planet Venus is near its extreme eastern or western elongation, no fixed star can bear the least comparison with it for splendour. When an evening star in the western sky, universal attention is always directed to it on account of the extra lustre it gives to that portion of the heavens. In the morning hours, however, Venus appears a still more magnificent object, owing principally to the clearer state of the atmosphere before sunrise. At the times of greatest brilliancy the light of Venus is very intense. A sensible shadow is often thrown upon a piece of white paper by the interposition of the hand between it and Venus when the planet is in this position of its orbit. It can also be plainly perceived by the naked eye at such times in full sunlight, sometimes within an hour of noon. At one of these epochs, in 1868, a correspondent of the *Times* fancied that he had discovered a balloon-shaped comet at noon-day by means of a small hand-telescope. The stranger, however, turned out to be the planet Venus, which happened to be favourably situated for daylight observation in the spring of that year.

Venus was named by the ancients Hesperus and Lucifer, or the evening and morning star, names by which it is frequently identified, especially by our poets. Jupiter shares with Venus the popular appellation of the morning and evening star, and when in an elevated position in winter its lustre rivals that of Venus. Milton speaks of the latter as the

"Fairest of stars, last in the train of night;
If better thou belong not to the dawn.
Sure pledge of day, that crown'st the smiling morn
With thy bright circlet, praise him in thy sphere,
While day arises, that sweet hour of prime."

Venus is nearly of the same magnitude as the Earth, its diameter being about 7,500 miles. The sidereal period, or time of revolution round the Sun, is 224^d 17^h, and the time of its rotation on its own axis is about 23^h 21^m. Its orbit, which deviates but little from a circle, is nearly midway between those of the Earth and Mercury, the mean distance from the Sun being 66,431,000 miles. When nearest the Earth at inferior conjunction Venus is 25 millions of miles from us. As a telescopic object this planet is far too brilliant to allow the markings on its surface to be distinctly seen; but some astronomers have perceived ill-defined dusky spots, from which the assumed time of rotation has been determined, although much doubt has been expressed on the permanency of the spots used by Cassini in 1667 for that purpose. Sir W. Herschel occasionally saw spots on the disk of Venus, but he could not bring his mind to believe that they were anything but optical delusions. "For," he remarks, "the spots assumed often the appearance of optical deceptions, such as might arise from prismatic affections, and I was always very unwilling to lay any stress upon the motion of spots that either were extremely faint and changeable, or whose situation could not be precisely ascertained." Many observers have, however, since scrutinised the planet very closely, including Schröter, Beer, Mädler, and De Vico, all of whom confirm in some measure the early observations of Cassini.

Viewed through a telescope, Venus exhibits phases of remarkable distinctness. Galileo, in 1610, was the first person who saw this phenomenon. It appears from a letter still in existence, that Father Castelli, a celebrated philosopher at Florence, asked Galileo if Venus and Mars ought not to present phases similar to the Moon. The idea does not appear to have crossed the inquiring mind of Galileo, or, if it had, he was cautious in propagating new astronomical facts, knowing full well the consequences, for Father Castelli received the following laconic reply:—"I am now occupied in so many researches, and the state of my health is so unsatisfactory, that I find myself far better lying in bed than exposing myself to the damp and chill of out of doors." Within six weeks, however, Galileo found time and inclination to explore the heavens with a telescope newly constructed, and was rewarded by the view of the crescent of Venus, which he announced to Castelli on December 30, 1610.

The telescopic appearance of Venus may be likened to a miniature Moon. In short, their phases are produced by causes perfectly similar, depending on their relative positions in their orbits, with respect to the Sun and Earth. Venus is therefore sometimes round, then gibbous, then like a half-moon, and finally a crescent, which directly before inferior conjunction becomes so narrow as to appear like a curved illuminated hair. The crescent form is exhibited from the time of inferior conjunction to that of greatest elongation, when the planet becomes half-illuminated. Between the greatest elongation and superior conjunction, it assumes a gibbous form, becoming quite round when in a direct line beyond the Sun. The apparent magnitude of Venus varies to a considerable amount in the course of its revolution in its orbit, as its distance from the Earth increases or diminishes.

Although Venus is now and then comparatively so near us, we know but little of its actual surface, principally owing to its intense brilliancy, which dazzles the eye of the observer. With regard to the two planets, Mercury and Venus, it is known that they are globes formed similarly to our Earth, and equally illuminated and warmed by the Sun. It is believed, also, from special observations of the physical appearance of the surface of Venus, that clouds prevail; if so, there must be water, and probably an atmosphere. This hypothesis would seem to be partially borne out by a phenomenon observed during the transit of Venus across the Sun's disk in 1761. While projected on the Sun, the planet appeared surrounded by a faint nebulous ring, and at the moment when Venus left the Sun, a luminous ring was observed on the Sun's edge. These two phenomena could be easily explained, if we suppose the globe of Venus to be surrounded by a very dense atmosphere. Farther than this, the most powerful instrument of the astronomer is unable to add to the little knowledge we possess of the actual formation of these intra-terrestrial planets. Of their peculiar motions in the heavens with respect to the fixed stars, and their effects on each other by their mutual attractions, the results obtained from modern astronomical observations leave but little more for us to learn.

Neither Mercury nor Venus has any satellite visible through the most powerful telescope yet constructed. If Venus had an attendant which could bear any comparison with the Moon, it would most probably have been detected long before this, by some of our modern astronomers. Some of the most vigilant observers have, under every favourable circumstance of atmosphere, watched the planet for this purpose, but without success. Little faith can therefore be placed in the observations of the elder Cassini, Short, Montaigne, Horrebow, Montbarron, and others, who have recorded their opinions that on several occasions they obtained a view of such a satellite. Professor Lambert even collected all the reputed observations, from which he succeeded in computing a tolerably consistent orbit, the period of revolution being rather more than eleven days, and the distance from Venus nearly the same as that of the Moon from the Earth. M. Guillemin remarks that " Venus is very similar to the Earth in many points, both in its dimensions and physical constitution. If we accept the great number of observations made in the seventeenth and eighteenth centuries, it would have still another point of resemblance. As the Moon accompanies the Earth, Venus would be also provided with a satellite. But this singular body cannot now be seen, and high scientific authorities have declared that the observers had been deceived by an optical illusion.

It must be acknowledged that the doubt which still exists on this account is at least very curious, and shows that in the domain of planetary astronomy there still remains a disputed point to be cleared up."

The celebrated astronomer, Christianus Huygens, made some remarks in the second half of the seventeenth century on the planet Venus, which exhibit very clearly that our knowledge of the physical constitution of its surface has very little increased since his day. "The Sun," he observes, "appears to the inhabitants of Venus by half larger in his diameter, and above twice in his circumference, than to us, and by consequence affords them but twice as much light and heat, so that they are nearer our temperature than Mercury. Their year is completed in seven and a half of our months. In the night, our Earth, when 'tis on the other side of the Sun from Venus, must needs seem

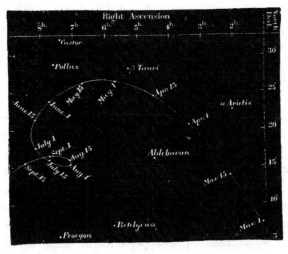

APPARENT PATH OF VENUS. 1868.

much larger and lighter to Venus than she doth ever to us; and then they may easily see, if they have not very weak eyes, our constant attendant the Moon. I have often wondered that when I have viewed Venus at her nearest to the Earth, when she resembled an half-moon, just beginning to have something like horns, through a telescope of forty-five or sixty feet long, she always appeared to me all over equally lucid, that I can't say I observed so much as one spot in her, though in Jupiter and Mars, which seem much less to us, they are very plainly perceived. For if Venus had any such thing as sea and land, the former must necessarily show much more obscure than the other, as any one may satisfy himself, that from a very high mountain will but look down upon our Earth. I thought that perhaps the too brisk light of Venus might be the occasion of this equal appearance; but when I used an eye-glass that was smoked for the purpose, it was still the same thing." To this testimony of the old astronomer, little can be added even now.

The change of the position of Venus with respect to the stars is very rapid when approaching to or receding from inferior conjunction. It may interest the reader to see the amount of this change at sight, which may be done by an inspection of the diagram. This diagram is intended to show the apparent motion of Venus in the heavens from March 1st to September 15th, 1868, and includes the period of greatest elongation, when the apparent motion is the minimum. In the interval of time included between the extreme dates, the planet has passed over more than a quarter of the heavens. The figures above the diagram represent the right ascension, or the distance in time from the vernal equinox, or, as it is technically called, the first point of Aries. From this point all angular distances in this direction, or right ascensions, are measured along the celestial equator. The figures at the side represent the declination, or the angular distance measured perpendicularly from the celestial equator. Right ascensions and declinations serve the same purpose for distinguishing the positions of celestial objects as longitudes and latitudes define the positions of places on a terrestrial globe or map.

THE EARTH.

The Earth, treated as a celestial globe, is a planet travelling in space in the same manner as the other members of the solar system. To the inhabitants, if any, of Venus or Mars it would appear a brilliant object in their skies, rivalling in lustre any other planet. From the Moon, it would appear to great advantage, casting so much light as thirteen Moons united. Its form is not a perfect sphere, the diameter from the north to the south pole being smaller than the diameter at the equator by about the three-hundredth part. A proof of the spherical form of the Earth is shown by the gradual appearance or disappearance of ships at sea. If we were to place ourselves on an elevated position near the sea, the first signs of an approaching vessel would be the tops of the masts, then the upper sails would gradually become visible, to be succeeded by the lower sails, and finally the hull would come into view. Again, if we are journeying on the ocean towards the southern hemisphere, we might daily watch the gradual declension of Polaris towards the north horizon, or if proceeding northwards, it would be observed as gradually increasing its altitude from day to day. Now this star, as we have before observed, is so near the true pole of the celestial sphere, that its position with respect to the zenith remains nearly the same throughout the year, if our station on the Earth be the same. If the Earth were not a globe, could it be for a moment supposed that the apparent declension or elevation was produced by a real increase or diminution in the absolute distance between us and the star? Such an argument would be positively absurd, because we know that the position of the fixed stars never varies sensibly, even when viewed from opposite sides of the Earth's orbit, a distance amounting to more than 183,000,000 miles. When, therefore, we observe Polaris, or any other star on the meridian, change its position in relation to the zenith and horizon, when viewed from different latitudes, but in the same longitude on the Earth's surface, the depression or elevation must be considered to arise solely from the curvature of our globe.

The great circle of the heavens, or the path which the Earth traverses in its revolution around the Sun, is called the ecliptic. This circle is inclined to the celestial equator, which is a projection of the terrestrial equator extended to the heavens, at an angle of about 23° 27′, termed generally the obliquity of the ecliptic. The two points of intersection of the celestial equator by the Earth's path, or by the Sun's apparent path, the positions of the two bodies being always separated by 180°, are called the equinoxes, while the two points on the ecliptic exactly 90° distant from

ILLUSTRATION OF THE ROTUNDITY OF THE EARTH.

the equinoxes, are called the solstices. Longitudes are reckoned along the ecliptic from the spring equinox, and right ascensions, as before explained, on the equator.

To this inclination of the Earth's axis we owe the variations of the seasons, for "while the Earth remaineth, seed-time and harvest, and cold and heat, and summer and winter, and day and night, shall not cease." As the direction of the axis of the Earth never changes, in whatever portion of its orbit it may be situated, the Earth passes onwards in its course with its axis pointing always to the same vanishing point in the sphere of the fixed stars, being carried around parallel to itself. As a consequence of the combination of this annual motion around the Sun with that produced by the rotation of the Earth on its axis, we have the division of the twenty-four hours into day and night, and the succession of the seasons. On the 21st of June, when the Sun is in the highest point of the ecliptic, the north or upper pole of the Earth is turned towards the Sun, when in all places north of the equator the longest day takes place, and in all places south of the equator the shortest day. On the 21st of December, the Earth being in exactly the opposite side of its orbit, but its axis remaining parallel to its position in June, it follows that the north or upper pole is now turned away from the Sun, while the south or lower pole is turned in that direction, and that the shortest day takes place at this time in the northern hemisphere, and the longest day in the southern hemisphere. At the equinoxes on March 21st and September 21st, the axis of the Earth is at right angles to the direction of the Sun, and there is equal day and night everywhere over the whole globe.

DIAGRAM OF THE SEASONS.

The dimensions of the Earth are known with considerable accuracy. We have previously mentioned that its form is not a perfect sphere, but that the poles are slightly compressed. It is more properly termed a spheroid. In a recent investigation by Captain A. R. Clarke, R.E., of the Ordnance Survey, he has deduced from a large number of measured arcs, that the form of the Earth is not even strictly a spheroid, but that it is an ellipsoid with three unequal axes, consisting of the polar axis, and two equatorial axes, differing from each other in length about two miles. From this Captain Clarke has inferred that the diameter of the Earth at the equator varies slightly in different longitudes. Assuming the Earth is a spheroid, the two best determinations of its magnitude are those of the Astronomer Royal and M. Bessel. The diameter in miles as given by these astronomers and by Captain Clarke, are :—

		Polar diameter.			Equatorial diameter.
Airy	. .	7899·17	. .		7925·65
Bessel	. .	7899·11	. .		7925·60
Clarke	. .	7899·16	. .		7926·70 in 14° E. long.
Do.	. . .	..	. .		7924·69 in 104° E. long.

The orbital velocity of the Earth equals 65,828 miles per hour, while the rate of speed at the equator depending upon the diurnal rotation, is about 1040 miles in the same interval. If the great centre of the system were divided into 1,262,000 equal parts, one of these parts would be of the same bulk as the Earth, but the density of the latter being four times greater than that of the Sun, four of these parts would be required to form a globe corresponding in weight to the Earth.

There are various methods of determining the weight of our globe. The subject of weighing the Earth has indeed been a favourite one, although from the difficulty experienced in the experiments, success has not always been the result of much care and labour. Among those who have made observations for the purpose are Dr. Maskelyne, Messrs. Cavendish, Reich, and Baily, the Astronomer Royal, and Colonel Sir Henry James, Director of the Ordnance Survey. No person has taken a greater interest in this subject than Mr. Airy, the Astronomer Royal, who by means of a series of pendulum observations on the surface, and at the bottom of a shaft of a deep mine, was enabled to observe the variation of the force of gravity acting on a free pendulum at the two stations. He made two unsuccessful attempts in 1826 and 1828 in Dolcoath mine, near Camborne, Cornwall; but in 1854, under much more favourable circumstances he planned a most elaborate series of observations, which were carried out in Harton coal-pit, near South Shields, by six skilled astronomical observers under the personal direction of the author. As this series of pendulum experiments for the determination of the density of the Earth is the most perfect ever made, a brief detailed account of the process may be instructive as well as interesting.

It may be mentioned at first that the result required from these experiments is the amount of the variation in the gravitational force acting on a pendulum at the top and bottom of the mine, and to assume that the increase of force at the bottom is owing to the greater density of the Earth in the interior than at the surface. When a pendulum is drawn away sideways by the hand, it will, as soon as it is released, return to its vertical position, which it will pass, to almost the same position on the other side. The force which makes this return is the force of gravity, or the attraction of the Earth pulling the bob downwards. Thus, at each vibration the gravitational force is exercised, until, from the density of the air and other causes, the pendulum, after an interval of some hours, is again at rest. This supposes, however, that the pendulum is suspended in such a manner that friction is avoided as much as possible. In the Harton experiments, two invariable pendulums were used, one of which was mounted in a building on the surface, and the other almost vertically, 1260 feet below. The upper station, which was formerly a stable near the mine-office, was fitted up expressly for the experiment, by the owner of the mine. The soft ground was removed to a considerable depth, and the space filled with square stones, and finally paved with great flat stones, to give a solid bearing to the apparatus; additional walls and roof were added, the spaces between them being filled with non-conducting materials, to prevent great changes in the temperature. Gas was laid on, so that the observer had the power of equalising the temperature if necessary. In the lower station, the floor of the solid rock was cut level and paved; brick walls were erected, a ceiling was put on; and in this manner three rooms were strongly made.

In each station a firm iron tripod-stand was erected. On this stand the pendulum was suspended by means of a projecting piece of hard steel, one end of which was ground to a knife-edge. This edge rested on planes of polished agate, fixed in a frame firmly attached to the tripod-stand. By this arrangement, very little friction could take place. Behind the pendulum, an ordinary astronomical clock was placed. A small inclined disk, covered with gold leaf and illuminated by a lamp, was fixed to the bob of the clock pendulum. This illuminated disk was viewed by means of a small telescope, attached to a block in an aperture in the wall of the pendulum room. Immediately in front of the gilt disk, and projecting from the bob of the free pendulum, a long narrow tail was suspended, the adjustment being made in such a manner that when the two pendulums were at rest, the bright disk was invisible. Now, if the clock pendulum were made to vibrate while the free pendulum was at rest, the disk, or its bob, would be visible alternately on the two sides of the pendulum tail. A pair of cheeks of thin metal was inserted into the opening of the clock-case, through which the disk was seen, the narrow slit being so arranged that when the clock pendulum was vibrating, the disk could not be seen on either side of the tail. The apparatus when in this condition was ready for use. If at this time the free pendulum were also made to vibrate, it can be easily perceived that when the two pendulums were not vibrating together, the illuminated disk was always visible when it passed the small aperture in the clock-case; but when the two pendulums were vibrating in coincidence, which always took place at short intervals, then the disk was invisible. Now, it was the business of the observers to note by the clock the exact second of each coincidence as shown by the disappearance of the disk from view, and by its reappearance a few seconds afterwards. By this operation, the number of vibrations made by the free pendulum in each set was easily ascertained. Before, however, the number of vibrations at the upper station could be compared with the number of vibrations made by the free pendulum at the lower station, a very important part of the experiment had to be made, by the comparison of the two clocks, with which the respective free pendulums were compared. In the Cornwall experiments, this comparison was made directly by means of chronometers carried up and down the mine, but in those at Harton, wires were carried down the shaft, connecting the two clocks, for the easy transmission of galvanic signals. The success of the observations in the present instance is owing in some measure to this precaution, for the comparisons were made as easily as if the two clocks were placed side by side.

The method of comparison was as follows:—A galvanometer, or ordinary telegraph needle, was placed near to each clock-face, in a convenient position for the observer to note the time of deflection of the needle. In the upper station, a battery of twenty-four cells was connected by insulated wires with the galvanometer. These wires passed through the works of an auxiliary time-piece, in which, by a contact made by pins fixed on one of the wheels, the galvanic current could pass through two springs, and then through the wires instantaneously to the coils of the two galvanometers. The auxiliary clock had the power of making a signal in this manner every fifteen seconds. At pre-arranged times, the two observers noted the clock-times at which the needles were deflected; and as the galvanic current performed its circuit instantly through the wires connecting the two stations, the difference between the upper and lower clocks was found with the greatest accuracy.

But the object of the experiment was to determine how much the free pendulum at the upper station had gained or lost on the free pendulum at the lower station. This comparison could, however, only be obtained indirectly by determining in the first place how much the free pendulum had, in each case, gained over the clock pendulum; and knowing, by the galvanic comparisons, how much one clock pendulum gained over the other, then the gain or loss, or in other words the excess of vibrations made by one free pendulum over the other, was easily determined. There were many minor, but necessary, precautions to be taken before a correct result could be obtained; for instance, the temperature of the air, the height of the barometer, etc., had to be registered at both stations, because the swings of the pendulum were considerably affected by the varying density and other conditions of the atmosphere, corrections for which had afterwards to be applied in the reductions.

The method of observation followed during the course of the experiments consisted of four independent series, the pendulums being interchanged at the end of each. During the first week twenty-six sets of vibrations were observed at each station, each set occupying about four hours, so that incessant observations were continued during 104 hours. Observations of the galvanic signals were always made at the beginning and end of each four-hourly set. Two of the six observers were engaged, one above and one below—the watch lasting twelve hours. The observations continued day and night. At the end of the first week, the two free pendulums, with parts of the apparatus which might have some influence on the results, were interchanged. During the second week, as in the preceding series, the observations were made incessantly from Monday morning to Friday evening. On the Saturday, the pendulums, etc., were interchanged again. In the third week, the interchange was made during the night of Wednesday, and on Saturday evening the final observation was made. These interchanges were necessary, to destroy any defect peculiar to either of the free pendulums, to the agate planes upon which they vibrated, or to the thermometers which exhibited the temperature of the surrounding air. It was intended also to interchange the iron stands, but when once fixed in position, their stability was so great that it was considered wiser not to interfere with them during the experiments.

These scientific operations were carried on whilst the labourers at the mine were in full work. Indeed, it was no unusual circumstance to see several of these hardy pitmen resign their position at the mouth of the shaft, to allow the " philosophers" to descend in the cage to their mysterious chamber, discussing, meanwhile, many different opinions on the object of this curious enterprise. During the three weeks occupied by the experiment, nothing untoward occurred to interrupt the observations; but on the day after the instruments had been removed, an accident occurred in the shaft to some of the lifting apparatus, which, had it happened in the course of the experiments, would have caused considerable delay, perhaps injuriously affecting the result.

It has been already stated that the direct object of these experiments was to ascertain the amount of the daily acceleration of a seconds' pendulum at the bottom of a mine, over another placed in similar circumstances on the surface; and to assume that any increased velocity attained at the lower station was caused by a greater force of gravity acting on the pendulum. Now, it was found from these experiments that this force of gravity was greater at the lower station than at the upper, by about $\frac{1}{10000}$ part of the whole force, and that its effect on the pendulum was such as to cause that at the lower station to make two and a quarter vibrations a day more than the corresponding one at the upper station.

Several other important considerations had to be taken into account before the final result could be obtained, giving the mean or average density of the Earth. For example, the formation of the crust or shell between the two stations had to be completely investigated; the surface of the neighbouring country was surveyed; and the thickness of every stratum of earth, rock, coal, etc., was measured in the shaft. One hundred and forty-two different layers were found. Of these, the specific gravity of the principal strata was determined by Professor Miller, of Cambridge. From these specimens, the average density of the Earth's crust at Harton is about two and a half times greater than water.

After combining the amount of daily acceleration of the lower pendulum, with the specific gravity, etc., of the Earth's surface, according to mathematical principles far too abstruse for these pages, the globe on which we dwell was found to have a mean density of six and a half times greater than that of water. This result is somewhat larger than those obtained from former experiments; but the Astronomer Royal considers that "it is entitled to compete with the others, on at least equal terms."

From these researches, it is not difficult to determine the actual weight of the Earth in pounds avoirdupois, knowing as we do the exact size and volume of our globe; but it is not the Earth only which can be weighed in this manner, for, using its mass and weight as the units of measurement, the masses and weights of the Sun and all the planets of the solar system can be inferred. That the knowledge of these masses is invaluable, it is only necessary to say, that the accurate predictions of the positions and motions of the heavenly bodies could never be performed without reliable data concerning the power of attraction which one body has on another, which power is regulated in a considerable degree by their masses.

Sometimes in the evenings of spring, near the equinox, a luminosity in the form of a cone may be observed shooting upwards from the horizon where the Sun had set, reaching a considerable distance towards the zenith. It has been frequently taken by those who are not familiar with the ordinary aspect of the sky, for the Milky Way, for an aurora, or the remains of twilight. It is generally known, however, by the name of the Zodiacal Light. It is observed much more favourably in tropical countries, where it has been noticed from day to day, and its position, with respect to the stars, recorded with the greatest minuteness. What this celestial phenomenon is composed of has only been a matter of conjecture. It has been suggested that it may probably proceed from the reflected light of myriads of meteors which are now known to be traversing space even to the extreme limits of the solar system. Humboldt, however, considers that it may possibly be a vast nebulous ring, rotating between the Earth and Mars, or, less probably, the exterior stratum of the solar atmosphere. He thus remarks on its appearance in the equatorial heavens :—"Those who have lived for many years in the zone of palms must retain a pleasing impression of the soft beauty with which the zodiacal light, shooting pyramidally upwards, illumines a part of the uniform length of tropical nights. I have seen it shine with an intensity of light equal to the Milky Way in Sagittarius, and that not only in the rare and dry atmosphere of the summits of the Andes at an elevation of from thirteen to fifteen thousand feet, but even on the boundless grassy plains, the Llanos of Venezuela, and on the seashore, beneath the very clear sky of Cumana. This phenomenon was often rendered especially beautiful by the passage of light fleecy clouds, which stood out in picturesque and bold relief from the luminous background. In our gloomy so-called 'temperate' northern zone, the zodiacal light is only distinctly visible in the beginning of spring, after the evening twilight, in the western part of the sky, and at the close of autumn, before the dawn of day, above the eastern horizon."

Other travellers have noticed the superior brilliancy of the zodiacal light when viewed from stations situated within the tropics. Major Tennant, R.E., made a series of observations of its position among the stars from the Bay of Bengal, during a voyage to Calcutta in the early part of 1868. He frequently watched it, at intervals from sunset until nine in the evening; and after the glow of sunset was gone he always found the shape to be a portion of a long ellipse, or parabola, ill-defined at the outlines, but towards the axis and horizon there was a marked condensation of light. M. Du Chaillu, in his African wanderings, was also much impressed with the beauty of this phenomenon. When occasionally enchanted by the superb appearance of the heavens, at a time when most of the constellations to which we have previously alluded have been only a few degrees south of the zenith, the zodiacal light has appeared to add fresh interest to the scene. Referring to the magnificent spectacle, he remarks :—"Then, as if to give a still grander view to the almost enchanting scene, the zodiacal light rose after the Sun had set, increasing in brilliancy, of a bright yellow colour, and rising in a pyramidal shape high into the sky, often so bright that it overshadowed the brightness of the Milky Way and the rays of the Moon, the beautiful yellow light gradually diminishing towards the apex. It cast a gentle radiance on the clouds round it, and sometimes formed almost a ring, but never perfect, having a break near the meridian; at times being reflected in the east with nearly as much brilliancy, if not as much, as in the west, and making me almost imagine a second sunrise. April and May were the months when the light showed itself in its greatest brilliancy. On April 13th, 1865, the glow coming from the west was so bright that it totally hid the Milky Way in the principal part of its course. I could only distinguish it above the sword of Orion; the glow was the brightest below the planet Mars, and the base of the pyramid reached, on the south, the part of the Milky Way, at the foot of the Cross." The zodiacal light, as observed in England, can bear no comparison in intensity with the appearances above, probably owing to the greater amount of twilight north of the tropical zone. It has, however, been seen in this country, very favourably on some occasions; at times we have been able to notice and record its relative position with respect to neighbouring celestial objects with great exactness. To an observant eye it is sometimes distinguishable in the spring evenings without much effort. But the equatorial skies alone must be scanned to see with advantage the pyramidal form and the superior lustre of this curious phenomenon.

M. Borelly, of the Marseilles Observatory, observed, on the evening of the 30th of January, 1869, a very splendid exhibition of the zodiacal light. While searching for comets in the western part of the sky, the field of his telescope was lightened up with it, so that the small stars were very difficult to see. The light mounted upwards from that part of the horizon near which the Sun had set.

MARS.

Mars is the fourth planet from the Sun, being the next in order from the Earth. Its surface exhibits a greater analogy to that of our own globe than that of any other planet of the solar system; and when it is at its least distance from us, it shines with great splendour in the midnight sky. Mars can always be distinguished from the other planets, and from the fixed stars, by its ruddy light. Owing to its occasional near approach to our Earth, great facilities are obtained for delineating the various lights and shades on its disk; and at such opportunities numerous accurate drawings are always made. When viewed through large telescopes, the surface of Mars represents the outline of continents and seas, and near the poles white spots are clearly visible, which owe their existence probably to an accumulation of snow in the polar regions.

The mean distance of Mars from the Sun is one hundred and forty millions of miles, and its nearest distance from the Earth averages forty-eight millions. The equatorial diameter of Mars is about 4,363 miles, and its polar diameter about seventy miles less. Mars revolves around the Sun in a few minutes less than 687 days. When in or near opposition, the form of Mars is sensibly globular, but in other portions of its orbit it is generally seen gibbous.

The apparent magnitude of Mars is very variable. When on the opposite side of the Sun with respect to the Earth, and at its greatest distance from us, its telescopic diameter amounts to less than four seconds of arc; but at the times of its nearest approach at favourable oppositions, its telescopic diameter is increased to twenty-four seconds of arc. When viewed on these occasions with a good telescope, the disk of Mars appears covered with various markings of a very distinct character, some of which give those defined appearances of continents and seas which have been so ably depicted by so many astronomers. The brightest parts, excepting the white patches near the pole, have a ruddy tint, while the darker portions have a greenish hue, the effect probably of contrast. It is by the observation, at different epochs, of the positions of these fixed markings on the disk of Mars, that the most accurate determinations of its rotation have been made. One recently published by Mr. Proctor gives the value $24^h 37^m 22 \cdot 73^s$, which is probably true to the hundredth part of a second.

If there be any inhabitants on Venus, the surface of the Earth will appear to them very similar to that exhibited to us in telescopic views of Mars. They will, however, have the advantage of seeing a globe of larger dimensions, but the general aspect of the terrestrial surface, including the distribution of land and water, will be the same. The poles of the Earth would probably appear comparatively bright, as in Mars, if it be true that our unexplored polar regions are covered with ice and snow. Although the atmosphere of Mars is not considered to be so dense as was formerly attributed to that planet, yet it is proved that, like the Earth, it is surrounded by one of sufficient density to obscure occasionally the various markings on its surface, particularly those near the edges of the planet. Mars is doubtless subject to similar meteorological phenomena as our Earth, and to perhaps greater sudden changes of weather. Professor Phillips, of Oxford, has remarked, that the great interchange of the humidity of the atmosphere which must necessarily take place periodically between the two hemispheres, and especially between the two poles, would give rise to very violent hurricanes, of which we have little conception. The difference in the lengths of the years of the Earth and Mars, is one point where the analogy between the two planets fails, for as the Martial year is nearly twice as long as that of the Earth, the seasons on Mars will be lengthened in a corresponding proportion. For example, in the northern hemisphere of Mars, spring lasts 191 Martial days, summer 181 days, autumn 149 days, and winter 147 days; and as the seasons are reversed in the southern hemisphere, spring and summer, taken together, last seventy-six days longer in the northern hemisphere than in the southern.

Mars is the only large planet, exterior to the Earth, without a satellite.   Being at a greater distance from the Sun than our globe, one would suppose that there would be a greater necessity for an attendant Moon than with us, but no telescopic aid has been able to detect any object near the planet.   If, however, a very small Moon were situated comparatively near to the surface of Mars, our present optical means would probably not be sufficiently powerful to perceive it.   In the absence of a satellite, the nights of Mars must be always obscured, relieved only by occasional auroral displays, or by the morning and evening twilight before and after the Sun is above the horizon. The mass of Mars is only one-eighth part of that of the Earth.   Its orbital velocity, or motion in space around the Sun, is 53,514 miles per hour; and the velocity of its rotation at the equator 558 miles per hour.

Until recent years the ruddy tint of Mars was universally believed to owe its origin to an unusually dense atmosphere; but this opinion has been considerably modified since the surface of the planet has been so carefully examined by Professor Phillips, Mr. De La Rue, Mr. Lockyer, Mr. Huggins, and others.   It has been found that the light reflected from the neighbourhood of the poles has no trace of colour, although in its course it has passed through a denser atmosphere than that which is found on the central portions of the disk, where the ruddy tint is most apparent.   Mr. Huggins, who has made some very important observations of the spectrum of the solar light reflected from Mars, remarks that "if indeed the colour be produced by the planet's atmosphere, it must be referred to peculiar conditions of it which exist only in connection with particular portions of the planetary surface.   The evidence we possess at present appears to support the opinion that the planet's distinctive colour has its origin in the material of which some parts of its surface are composed.   Mr. Lockyer's observation, that the colour is most intense when the planet's atmosphere is free from clouds, obviously admits of an interpretation in accordance with this view."

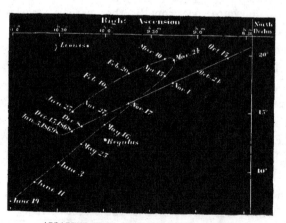

This diagram of the path of Mars indicates the change of the position of the planet from October, 1868, to June, 1869, and exhibits, as an example, its retrograde motion among the stars as viewed from the Earth. Its decreasing right ascension was very evident to the naked eye during the first three months of 1869, by comparing its position, at intervals of a few days, with the principal stars in the well known Sickle in the constellation Leo.

APPARENT PATH OF MARS, 1868 AND 1869.

THE MINOR PLANETS.

AS soon as the relative positions of the planets in the solar system became approximately known, it was observed that a remarkable harmony existed in the progression of their distances from the Sun, but that its uniformity was broken between Mars and Jupiter. This regularity was first pointed out by Kepler, more than three hundred years ago, but it was not till after the discovery of Uranus, in 1781, that any particular attention was directed to this supposed law, when Professor Bode, of Berlin, published a modification of it, now known as Bode's "empirical law of planetary distances." According to this law there ought to be a planet, or something equivalent to one, revolving around the Sun in an orbit situated between those of Mars and Jupiter; and, to detect such a body, an association of astronomers, under the guidance of the Baron de Zach, of Gotha, was formed for the express purpose of making a strict and systematic search for the supposed unknown member of the solar system. This examination of the heavens within a limited space on each side of the ecliptic, was prosecuted with great zeal and energy in the closing years of the last century. Great was the excitement, therefore, when Professor Piazzi, of Palermo, announced that on the first day of the new century, January 1st, 1801, he noticed a small object, which, by its motion relatively to the neighbouring stars, he considered to be a comet, but which Professor Gauss, by computing its orbit, found to be a planet, named afterwards Ceres. This discovery was followed by three others, Pallas, Juno, and Vesta, the last in 1807. The search was continued during several years, especially by Dr. Olbers, but without any further success.

    The discovery of these small planets confirmed in a great measure the accuracy of Bode's law: yet the existence of four instead of one, all at an unequal distance from the Sun, was not what that astronomer expected. Dr. Olbers started an hypothesis that Ceres and Pallas were probably fragments of a large one which had, by some catastrophe, been broken to pieces; and that it was very likely that other fragments, too minute to be detected, were revolving in similar orbits. When Juno and Vesta were added to the group, he considered that fresh evidence was gathered to give plausibility to his conjectures. The investigations of the motions of the numerous minor planets discovered in later years, do not, however, tend to confirm Dr. Olbers' theory.

The fifth of the minor planets was detected by M. Hencke, at Driessen, Germany, on the 8th of December, 1845. In 1846, the most distant of the planets, Neptune, was discovered. Since that date, no year has passed away without an addition to the known members of the solar system. The number of these minute planets between Mars and Jupiter now (September, 1869) amounts to *one hundred and eight*. Their orbits are all computed with tolerable accuracy, and systematic observations continue to be made of each when in favourable positions. Many circumstances connected with the discovery of some have been peculiarly interesting. It would occupy, however, too much space to give a detailed description of the different members of this group, further than to state that the majority of them were found by comparing limited portions of the heavens with either published or manuscript star-charts containing all stars down to the tenth or eleventh magnitude. A few were found accidentally while observing, or seeking for another small planet. Dr. Luther, of Bilk, near Dusseldorf, has discovered *seventeen*, the late M. Goldschmidt *fourteen*, and Mr. Hind *ten*. The following list contains all the information concerning them likely to be of any popular interest:—

NO.	NAME.	DATE OF DISCOVERY.	DISCOVERER.	PLACE OF DISCOVERY.
1	Ceres.	1801, January 1.	Piazzi.	Palermo.
2	Pallas.	1802, March 28.	Olbers.	Bremen.
3	Juno.	1804, September 1.	Harding.	Lilienthal.
4	Vesta.	1807, March 29.	Olbers.	Bremen.
5	Astræa.	1845, December 8.	Hencke.	Driessen.
6	Hebe.	1847, July 1.	Hencke.	Driessen.
7	Iris.	,, August 13.	Hind.	London.
8	Flora.	,, October 18.	Hind.	London.
9	Metis.	1848, April 25.	Graham.	Markree.
10	Hygeia.	1849, April 12.	De Gasparis.	Naples.
11	Parthenope.	1850, May 11.	De Gasparis.	Naples.
12	Victoria.	,, September 13.	Hind.	London.
13	Egeria.	,, November 2.	De Gasparis.	Naples.
14	Irene.	1851, May 19.	Hind.	London.
15	Eunomia.	,, July 29.	De Gasparis.	Naples.
16	Psyche.	1852, March 17.	De Gasparis.	Naples.
17	Thetis.	,, April 17.	Luther.	Bilk.
18	Melpomene.	,, June 24.	Hind.	London.
19	Fortuna.	,, August 22.	Hind.	London.
20	Massilia.	,, September 19.	De Gasparis.	Naples.
21	Lutetia.	,, November 15.	Goldschmidt.	Paris.
22	Calliope.	,, November 16.	Hind.	London.

NO.	NAME.	DATE OF DISCOVERY.	DISCOVERER.	PLACE OF DISCOVERY.
23	Thalia.	1852, December 15.	Hind.	London.
24	Themis.	1853, April 6.	De Gasparis.	Naples.
25	Phocea.	,, April 6.	Chacornac.	Marseilles.
26	Proserpine.	,, May 5.	Luther.	Bilk.
27	Euterpe.	,, November 8.	Hind.	London.
28	Bellona.	1854, March 1.	Luther.	Bilk.
29	Amphitrite.	,, March 1.	Marth.	London.
30	Urania.	,, July 22.	Hind.	London.
31	Euphrosyne.	,, September 1.	Ferguson.	Washington.
32	Pomona.	,, October 26.	Goldschmidt.	Paris.
33	Polyhymnia.	,, October 28.	Chacornac.	Paris.
34	Circe.	1855, April 6.	Chacornac.	Paris.
35	Leucothea.	,, April 19.	Luther.	Bilk.
36	Atalanta.	,, October 5.	Goldschmidt.	Paris.
37	Fides.	,, October 5.	Luther.	Bilk.
38	Leda.	1856, January 12.	Chacornac.	Paris.
39	Lætitia.	,, February 8.	Chacornac.	Paris.
40	Harmonia.	,, March 31.	Goldschmidt.	Paris.
41	Daphne.	,, May 22.	Goldschmidt.	Paris.
42	Isis.	,, May 23.	Pogson.	Oxford.
43	Ariadne.	1857, April 15.	Pogson.	Oxford.
44	Nysa.	,, May 27.	Goldschmidt.	Paris.
45	Eugenia.	,, June 28.	Goldschmidt.	Paris.
46	Hestia.	,, August 16.	Pogson.	Oxford.
47	Aglaia.	,, September 15.	Luther.	Bilk.
48	Doris.	,, September 19.	Goldschmidt.	Paris.
49	Pales.	,, September 19.	Goldschmidt.	Paris.
50	Virginia.	,, October 4.	Ferguson.	Washington.
51	Nemausa.	1858, January 22.	Laurent.	Nismes.
52	Europa.	,, February 6.	Goldschmidt.	Paris.
53	Calypso.	,, April 4.	Luther.	Bilk.
54	Alexandra.	,, September 10.	Goldschmidt.	Paris.
55	Pandora.	,, September 10.	Searle.	Albany, U.S.
56	Melete.	1857, September 9.	Goldschmidt.	Paris.
57	Mnemosyne.	1859, September 22.	Luther.	Bilk.
58	Concordia.	1860, March 24.	Luther.	Bilk.
59	Olympia.	,, September 12.	Chacornac.	Paris.
60	Echo.	,, September 15.	Ferguson.	Washington.
61	Danaë.	,, September 19.	Goldschmidt.	Chatillon-sous-Bagneux.
62	Erato.	,, October 10.	Förster.	Berlin.
63	Ausonia.	1861, February 10.	De Gasparis.	Naples.
64	Angelina.	,, March 4.	Tempel.	Marseilles.
65	Cybele.	,, March 8.	Tempel.	Marseilles.
66	Maia.	,, April 9.	Tuttle.	Cambridge, U.S.
67	Asia.	,, April 17.	Pogson.	Madras.
68	Leto.	,, April 29.	Luther.	Bilk.
69	Hesperia.	,, April 29.	Schiaparelli.	Milan.
70	Panopea.	,, May 5.	Goldschmidt.	Fontenay-aux-Roses.
71	Niobe.	,, August 13.	Luther.	Bilk.
72	Feronia.	,, May 29.	Peters.	Clinton, U.S.
73	Clytie.	1862, April 7.	Tuttle.	Cambridge, U.S.
74	Galatea.	,, August 29.	Tempel.	Marseilles.
75	Eurydice.	,, September 22.	Peters.	Clinton, U.S.
76	Freia.	,, October 21.	D'Arrest.	Copenhagen.
77	Frigga.	,, November 12.	Peters.	Clinton, U.S.

NO.	NAME.	DATE OF DISCOVERY.	DISCOVERER.	PLACE OF DISCOVERY.
78	Diana.	1863, March 15.	Luther.	Bilk.
79	Eurynome.	,, September 14.	Watson.	Ann Arbor, U.S.
80	Sappho.	1864, May 3.	Pogson.	Madras.
81	Terpsichore.	,, September 30.	Tempel.	Marseilles.
82	Alcmene.	,, November 27.	Luther.	Bilk.
83	Beatrix.	1865, April 26.	De Gasparis.	Naples.
84	Clio.	,, August 25.	Luther.	Bilk.
85	Io.	,, September 19.	Peters.	Clinton, U.S.
86	Semele.	1866, January 6.	Tietjen.	Berlin.
87	Sylvia.	,, May 16.	Pogson.	Madras.
88	Thisbe.	,, June 15.	Peters.	Clinton, U.S.
89	Julia.	,, August 6.	Stéphan.	Marseilles.
90	Antiope.	,, October 1.	Luther.	Bilk.
91	Ægina.	,, November 4.	Stéphan.	Marseilles.
92	Undina.	1867, July 7.	Peters.	Clinton, U.S.
93	Minerva.	,, August 24.	Watson.	Ann Arbor, U.S.
94	Aurora.	,, September 6.	Watson.	Ann Arbor, U.S.
95	Arethusa.	,, November 23.	Luther.	Bilk.
96	Ægle.	1868, February 17.	Coggia.	Marseilles.
97	Clotho.	,, February 17.	Tempel.	Marseilles.
98	Ianthe.	,, April 18.	Peters.	Clinton, U.S.
99	Dike.	,, May 29.	Borelly.	Marseilles.
100	Hecate.	,, July 11.	Watson.	Ann Arbor, U.S.
101	Helena.	,, August 16.	Watson.	Ann Arbor, U.S
102	Miriam.	,, August 22.	Peters.	Clinton, U.S.
103	Hera.	,, September 7.	Watson.	Ann Arbor, U.S.
104	Clymene.	,, September 13.	Watson.	Ann Arbor, U.S.
105	Artemis.	,, September 16.	Watson.	Ann Arbor, U.S.
106	Dione.	,, October 10.	Watson.	Ann Arbor, U.S.
107	Camilla.	,, November 17.	Pogson.	Madras.
108	Hecuba.	1869, April 2.	Luther.	Bilk.

The nearest of the minor planets to the Sun is Flora, at a mean distance of 200 millions of miles. Cybele and Sylvia, the farthest from the Sun, are more than 300 millions of miles distant, the intermediate space being occupied by the group, the orbits of the different members of which are separated on the average by about one million of miles only. The actual linear dimensions of the minor planets are very variable; the largest of them, Ceres and Vesta, being, according to Sir W. Herschel, about 200 miles in diameter. Atalanta, Echo, and Maia, three of the smallest, are supposed to be less than twenty miles in diameter.

Melete, No. 56, was discovered by M. Goldschmidt on September 9th, 1857, whilst searching for Daphne, for which it was taken. Its order in the preceding table is regulated by the date when it was first identified by M. Schubert as a separate planet. Feronia was also observed during several months as Maia, till its independent existence was detected by Mr. Safford from a discussion of the observations.

THE MAJOR PLANETS.

"Next, belted JUPITER far distant gleams,
Scarcely enlightened with the solar beams.
With four unfixed receptacles of light,
He towers majestic through the spacious height;
But farther yet the tardy SATURN lags,
And eight attendant luminaries drags;
Investing with a double ring his pace,
He travels through immensity of space.
Next, see URANUS wheeling wide his round
Of fourscore years; not unassisted found
By human eye; the telescope displays
Him, with four moons, to philosophic gaze.
Still more remote, pale NEPTUNE wends his way;
The student's skill divined his distant ray.
His lengthened year, by his slow-moving pace,
A hundred sixty-four of ours may trace."

CHATTERTON (*amplified*).

JUPITER.

JUPITER is the largest planet of the solar system, and, omitting the group of minor planets, the fifth in order from the Sun. Its diameter is about 85,000 miles, and its bulk is nearly 1250 times that of the Earth. Jupiter is accompanied in its orbit by four moons visible with slight optical aid, and its system bears a complete analogy to that of which it is a member, obeying the same laws, and exhibiting in the most attractive manner the prevalence of the law of gravitation as the guiding principle of the motion of the satellites around their primary. The time occupied by a complete revolution of Jupiter round the Sun is nearly twelve years. Its average distance from the Sun is 478 millions of miles. Some idea of the extent of this interval of celestial space may be gathered from the fact that a cannon-ball, going at the rate of 500 miles an hour, would take more than ninety years to perform its journey between Jupiter and the Earth; or a railway steam-engine, travelling fifty miles an hour, would require nine centuries to pass over a like distance.

Jupiter revolves on its axis in about nine hours and fifty-five minutes; a Jovian day is therefore less than ten of our hours. Its mass, or weight, is 300 times greater than

271

that of the Earth; but as its bulk, or volume, is nearly 1,250 times greater, it follows that its density can only be one-quarter that of the Earth. Jupiter is passing through space at the rate of 28,743 miles an hour, and is also performing the equatorial revolution on its own axis at the rate of 27,726 miles an hour. As seen from the Earth, Jupiter does not present any sensible phase in ordinary telescopes, owing to its great distance from the Sun; but when observed through a powerful telescope the right or left edge of its disk shows occasionally considerable signs of want of illumination.

When viewed with the naked eye, Jupiter shares with Venus that universal attention which is always given to the evening and morning stars; but sometimes Jupiter shines with even greater splendour than Venus, especially when it is due south at midnight in the winter months. At these times, Jupiter passes the meridian at an altitude equal to that of the Sun in summer, while the light of Venus is frequently partially eclipsed by the twilight, or by the hazy nature of the atmosphere near the horizon. But when Venus is at its greatest brilliancy, the greater intensity of its reflected light makes it invariably the brighter planet of the two, although its diameter is much smaller than that of Jupiter. It is, however, as a telescopic object that Jupiter has become so popular and interesting to the astronomer, for by the application to the eye of a very ordinary telescope, the four attendant satellites or moons, and the distinctive lineaments of light and shade on its surface, become plainly visible. The motions of the satellites around Jupiter are very soon perceptible, as they are continually changing their positions with respect to the body of the planet. Sometimes they are seen to disappear into the shadow of Jupiter, and thus become totally eclipsed, similarly to our own Moon; at other times they are observed to pass behind the planet, and then reappear on the opposite side; and again at other times they may be noticed on the disk of the planet. This last appearance is a very interesting phenomenon, as not only is the image of the satellite projected on the disk, but its shadow also is generally seen at the same time as a small round black spot. On some occasions, Jupiter is seen apparently without satellites, which at these times are either obscured in the shadow, or projected on the disk of the planet, but this occurrence is very rare. The last phenomenon of this kind took place on August 21st, 1867, when, notwithstanding the general unfavourable state of the weather, some very interesting observations were made. The most curious was the appearance of the fourth satellite on the disk of Jupiter as a dark object, nearly as black as its shadow. From this observation, it has been inferred that the reflective power of this satellite, which is the most distant from Jupiter, must be greatly inferior to the other three, and that it is also of less intrinsic brightness than the body of the planet. The telescope also reveals to us that the surface of Jupiter is partially covered with brownish-grey streaks parallel with the equator. Two of these are very conspicuous, one north, the other south of the equator. They extend completely around the ball of the planet, for no great deviation in their form can be observed on opposite sides. These streaks, or belts, resemble in some measure the lines of stratus cloud often seen on calm evenings near the horizon about the time of sunset. Between the two principal belts, a more brilliant ground marks the equatorial region of the planet. Towards the poles, a continuation of parallel belts of different intensities can be observed. The illumination of the disk near the poles is evidently more feeble than near the equator. Occasionally dark round spots have been seen on the principal belts, which have afforded a good means for the determination of the time of rotation of the planet. Some excellent drawings of Jupiter have been made by Mr. De La Rue, Sir John Herschel, M. Mädler, and others.

The satellites of Jupiter were discovered by Galileo on January 7, 1610. He first imagined them to be fixed stars, in no way connected with the planet. By a mere accident he happened to examine Jupiter on the succeeding day, when he was surprised to find that the arrangement of the stars was quite different, three of them being now west of the planet, whereas on the preceding day two were on the east side, and one only on the west. On January 13, he saw four stars, when he came to the conclusion that they were small planets revolving around Jupiter in the same manner as the ordinary planets revolve around the Sun, forming, in fact, a miniature of the solar system. In the same way as they were first seen by Galileo through one of the newly-invented telescopes, so they can now be easily discerned through any ordinary instrument fitted with an astronomical eye-piece. The satellites are generally known by simple numbers in the order of distance from their primary, but they have been sometimes called Io, Europa, Ganymede, and Callisto. The periods of their revolution around Jupiter are:—

First satellite	1^d	18^h	28^m
Second ,, ...	3	13	14
Third ,, ...	7	3	43
Fourth ,, ...	16	16	32

The following diagram exhibits the apparent path of Jupiter from April 15th, 1868, to the end of February, 1869, including the period of retrograde motion in the Autumn of 1868 :—

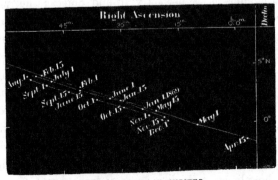

APPARENT PATH OF JUPITER.

SATURN.

Saturn is the sixth large planet from the Sun, and is one of the most beautiful celestial objects which come under the notice of the star-gazer. Saturn is a stupendous globe, about seven hundred times greater in volume than the Earth, surrounded by a series of rings of solid matter, and accompanied in its course around the Sun by eight satellites. All this complicated system moves with a common motion so exact that no part interferes with another in their orbital revolution, which is performed in rather less than thirty years. The mean distance of Saturn from the Sun is more than nine times that of the Earth, and nearly double the distance of Jupiter. The equatorial diameter of the ball is 74,000 miles, while its polar diameter is about 66,000 miles; its form is therefore sensibly elliptical, and is something the shape of a well-flattened orange. Streaks of light and shade have been observed on the ball, very similar to the better known belts of Jupiter, affording some kind of evidence of the existence of currents of air analogous to our trade winds. This supposition would lead us to infer that the surface of Saturn is surrounded by an atmosphere subject to all its attendant meteorological phenomena. Saturn revolves on its axis in 10^h 29^m 16^s, consequently a Saturnian day is less than half a terrestrial day.

But the system of rings which encircle the central ball of Saturn is by far the most interesting appendage of this magnificent planet. Before the invention of the telescope, the existence of the rings was unknown. Even the veteran Galileo was unable to view them satisfactorily, owing to the low penetrating power of the telescopes of his time. According to M. Arago, this low-defining power of Galileo's telescopes was a source of great perplexity to him. In a letter to the Grand Duke of Tuscany, he announced that the planet was "three-bodied," remarking that when he observed Saturn with a telescope magnifying thirty times, the central object appeared the greatest, the two others being attached, one on the east, and the other on the west side of the principal ball. At a later period Galileo observed Saturn at the time of one of its equinoxes, when the plane of the ring passes through the Sun, which then illuminates only the edge of it. On this occasion Saturn appeared to the illustrious astronomer as a perfectly round object; he was therefore led to conclude that his previous observations were nothing more than optical illusions. Galileo died in 1642, and it was not till Huygens observed the rings, in 1659, that a true explanation of the cause of the phenomenon was given. The gradual improvement in the construction of telescopes soon, however, supplied the means for observing the form and dimensions of the two bright rings as we see them at the present day, excepting only that the superior definition given by modern telescopes has enabled us to make further discoveries of great importance. For instance, the exterior ring, or that farthest from the planet, is seen separated from the intermediate ring by an empty space, showing that they are quite independent of each other. Between these and the ball of the planet, a dusky, or semi-transparent ring has been observed since 1850. The tint, or colour of the rings, is varied, the intermediate ring being the most brilliant of the three. The two exterior rings are opaque, and cast a very decided shadow on the disk of Saturn; but the dusky ring is evidently of a gaseous nature, as a part of the luminous disk can be seen through it. The thickness of the rings is very minute, so much so indeed, that when the edge is precisely directed to the Sun, it is nearly invisible even in telescopes of great power. It is assumed from this that the thickness cannot be much greater than 200 miles. At these times of its equinox, Saturn has the same globular, or rather spheroidal, appearance as the other large planets; and advantage is always taken of the invisibility of the rings to obtain accurate measures of the form of the central ball. The phenomenon of the disappearance of the rings of Saturn takes place at intervals of about fifteen years.

The mass, or weight of Saturn, is about ninety times greater than that of the Earth. The density is, however, only about the eighth part of the Earth's, or, in more

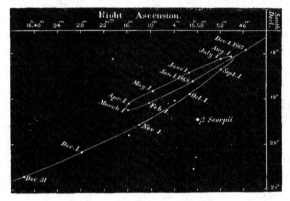

APPARENT PATH OF SATURN IN 1868.

familiar words, a cubic foot of the material of Saturn is eight times lighter than a cubic foot of the material of which the Earth is composed. Saturn moves in its orbit around the Sun at the rate of 21,220 miles per hour, and also turns on its axis at the equator 22,216 miles in the same time. It is impossible for any one to view this magnificent object through a large telescope without admiration, especially when the planet is in such a position with respect to the Sun and Earth as to exhibit the fullest extent of the rings. Many observers have, on such occasions, made elaborate drawings, showing all the minute peculiarities of the planet—those by Mr. Warren De La Rue and Mr. Dawes being almost perfect delineations of the disk and rings. Of the eight moons of Saturn, five were discovered between 1655 and 1684, two by Sir William Herschel in 1789, and one by Messrs. Lassell and Bond, in 1848. The satellites of Saturn are distinguished by the names of Titan, Japetus, Rhea, Dione, Tethys, Enceladus, Mimas, and Hyperion. The three last are very faint objects, visible only occasionally in our

275

largest telescopes. It has been remarked by Sir John Herschel, that at the time Enceladus and Mimas were discovered by his father, "they were seen to thread, like beads, the almost infinitely thin fibre of light to which the ring, then seen edgeways, was reduced, and for a short time to advance off it at either end, speedily to return, and hastening to their habitual concealment behind the body."

URANUS.

Uranus is the next planet in order of distance from the Sun. Like Jupiter and Saturn, it is the centre of a system, having at least four satellites, while it has probably more, but too small to be detected in our telescopes. The discovery of this large planet by Sir William Herschel in 1781, March 13th, was to some extent accidental. On the evening of that day the illustrious astronomer was examining the small stars in the neighbourhood of H. Geminorum, when he noticed one which seemed much lighter than the rest. The magnitude of this object was so much greater than any which he expected to find there, that his suspicions were aroused; and after watching it for some time he concluded that the stranger was a comet. Being engaged at the time in a series of observations on the parallax of the fixed stars, he had near him several eye-pieces of high magnifying power, with which he examined the new object. "I put on," he remarks, "two powers, 460 and 932, and found the diameter of the comet increased in proportion to the power, as it ought to be, on a supposition of its not being a fixed star, while the diameter of the stars to which I compared it was not increased in the same ratio. Moreover, the comet being magnified much beyond what its light would admit of, appeared hazy and ill-defined with these great powers, while the stars presented that lustre and distinctness which, from many thousand observations, I knew they would retain."

Observations of the supposed comet were made, not only by Herschel, but by all the principal astronomers of Europe. Owing, however, to its slow motion in the heavens, several months elapsed before its planetary nature was even conjectured, when it was provisionally placed in the solar system exterior to Saturn. It was not until Laplace communicated to the Academy of Sciences in January, 1783, elliptic elements of the planet, that all doubt was removed. Herschel, as the discoverer, gave the new planet the name of Georgium Sidus, but the French astronomers resolved to know it only by the name of Herschel. Professor Bode proposed Uranus, which was at first unpopular, but this name has been gradually superseding the other two, until, by universal consent, it is now adopted by the astronomers of all countries.

As a telescopic object, Uranus appears with a small round disk shining with a pale yellowish light. It is accompanied in its journey round the Sun by several satellites, but owing to its immense distance from us they are only seen, even in the best of telescopes, with the greatest difficulty. Herschel has recorded observations of six, but some of them have never been confirmed by later astronomers. In fact, we are only certain of four, two of which were observed by Herschel. The remaining two are nearer the planet than the innermost of Herschel's. These four satellites are known by the names of Ariel, Umbriel, Titania, and Oberon.

The diameter of Uranus is about 33,000 miles, and its volume or bulk is seventy-four times greater than that of the Earth. The solar light and heat reflected on the surface of a planet so distant as Uranus must be very small comparatively to what we are receiving, if a constant absorption of the light and heat has taken place while passing through space. Assuming the solar influence on the Earth to equal 1, that at Uranus, which is more than nineteen times farther from the Sun, amounts only to 0·003, or the three-thousandth part of that to which we are accustomed.

NEPTUNE.

The history of the discovery of Neptune is a record of one of the greatest triumphs of intellectual skill which have graced the progress of astronomical knowledge from the earliest ages to the present time. Two mathematicians, working independently of each other, each believing faithfully in the truth of the laws of universal gravitation, and having for the subject of their investigations certain observed irregularities in the motion of Uranus in its orbit, which could only be accounted for by supposing them caused by the attraction of some unknown large planet at a still greater distance from the Sun, announced to the practical astronomer, that if the adopted theory of the universe be true, the supposed body must be in a certain position of the heavens. These two mathematicians, M. Le Verrier, of Paris, and Mr. J. C. Adams, of Cambridge, sent the results of their calculations to the principal astronomers, some of whom organised a series of observations for a systematic search. When Dr. Galle, of Berlin, received on September 23rd, 1846, the indication of the place of the supposed planet from M. Le Verrier, he took advantage of a brilliantly clear evening which fortunately occurred on the same day, to direct his telescope to the spot communicated to him, when he soon noticed an object which was not inserted in the star-chart with which he was comparing the sky. He saw in about three hours that it had certainly moved, but on the succeeding evening he noticed that it had retrograded relatively to the neighbouring stars full four seconds of time. He then announced the discovery of the planet now known as Neptune, which has since been found to move in an orbit exterior to Uranus at a mean distance from the Sun of 2,760 millions of miles.

Although the actual discovery of Neptune resulted from the calculations of M. Le Verrier, yet it was partly by accident that it was so. Mr. Adams had placed his approximate elements of the disturbing planet in the hands of the Astronomer Royal and Professor Challis many months previously; and the latter had even commenced a search for the planet on the 29th of July, 1846. As the star-chart used by Dr. Galle was unknown in England, Professor Challis was obliged to make his own maps. Unfortunately, these manuscript maps were only partially compared to show whether

277

the plan of observation was effectual or not, for Professor Challis appears to have had no sanguine hopes at the time that "the indications of theory were accurate enough to give a chance of discovery in so short a time." After Dr. Galle's announcement, it was found that the Cambridge astronomer had actually noted the position of the planet in the map of the stars made on August 12th, and that when the previous partial examination had been made, the comparison was suspended at star No. 39, when the planet was No. 49 on the map.

The diameter of Neptune is reckoned to be about 37,000 miles, but owing to the remoteness of the planet, this value is subject to some uncertainty. Its mass is about the 19,000th part of that of the Sun, and the density of its material is about two-thirds of the density of the matter composing that body. On the surface of Neptune the Sun appears about the same magnitude as Venus when at its greatest brilliancy as viewed from the Earth; but the intensity of the Sun's light would be more than 10,000 times greater than that of Venus. As some compensation for the comparative absence of solar light, it is very probable that Neptune is furnished with several moons, in like manner with the other major planets. Our present telescopic means, however, have not sufficient space-penetrating power to enable us to detect many such minute objects at so great a distance. One satellite has undoubtedly been seen, first by Mr. Lassell, with his great reflecting telescope, in October, 1846, soon after the discovery of the planet. It was afterwards seen by M. Otto Struve, at Pulkowa, and by Professor Bond, of Cambridge, U.S., through the large equatorials of these two observatories. These instruments are the largest at present mounted, the object-glasses in each having a clear aperture of fifteen inches, the focal length being twenty-five feet. At one time a suspicion of a second satellite, and of an appearance similar to the rings of Saturn, was noticed by several astronomers, but it has not been confirmed. Mr. Lassell, who carried his superb twenty-foot reflector to Malta, to take advantage of the clear skies of southern Europe for deciding some of these delicate points of astronomy, has since declared his conviction that we have no authority to say that Neptune has more than one satellite, or that the supposed ring has any existence whatever.

Neptune and Uranus have been several times observed as fixed stars. Uranus was seen at the Royal Observatory so far back as the time of Flamsteed, who recorded its position at least on six occasions, not knowing at the time that it was a planet. Dr. Bradley, at Greenwich, also saw it in 1748 and 1750. Neptune was unconsciously observed by Lalande in 1795, and by Dr. Lamont in 1845 and 1846, before its discovery was announced by Dr. Galle in the latter year. As viewed with the transit-circle at Greenwich, Neptune appears as large as an ordinary star of the seventh and a half magnitude. The intensity of its light is however much less than that of a star of corresponding size.

COMETS.

——◆——

"Hast thou ne'er seen the Comet's flaming flight ?
Th' illustrious stranger passing terror sheds
On gazing nations ; from his fiery train,
Of length enormous, takes his ample round,
Through depths of ether ; coasts unnumbered worlds
Of more than solar glory ; doubles wide
Heaven's mighty cape ; and then revisits Earth
From the long travel of a thousand years."

YOUNG.

——◆——

FROM the earliest ages, comets have been regarded with a superstitious awe which has never been attached to the planetary members of the solar system. Louis the First of France was so much alarmed by the appearance of one which became visible in the year 837, that he felt it necessary to satisfy his mind and to appease the wrath of Heaven by ordering the construction of a number of churches and monasteries, which his conscience told him he had hitherto neglected. It is related by Gibbon that "in the fifth year of the reign of Justinian, A.D. 531, and in the month of September, a comet was seen during twenty days in the western quarter of the heavens, and which shot its rays into the north. Eight years afterwards, while the Sun was in Capricornus, another comet appeared to follow in Sagittarius ; the size was gradually increasing ; the head was in the east, the tail in the west, and it remained visible about forty days. The nations, who gazed with astonishment, expected wars and calamities from their baleful influence ; and these expectations were abundantly fulfilled." In 1456, at a time when Christianity was struggling against Mohammedanism in Eastern Europe, the appearance of a great comet, which has been since identified as that of the celebrated periodical comet of Halley, excited the utmost consternation in Rome, where it was regarded as a portentous sign of approaching evils. Pope Calixtus III. thereupon ordered prayers to be said in the churches, and the bells to be rung daily at noon to call upon the inhabitants to supplicate the Divine protection during the impending calamity. The progress of education since the time of the Reformation in Europe has very nearly banished these superstitious fears from the minds of all classes. These occasional visitors to our skies are now so universally

279

known to appear to us from natural causes, easily explained, that even the most brilliant of its class may be visible for months—the comet of 1811 for example—and yet the only excitement that may be noticed in the popular mind is that of extreme pleasure and delight at witnessing the progress of the wanderer across the heavens.

Comets, next to meteors, are the most numerous of all the bodies which revolve around the Sun. Mr. Hind supposes that, since the Christian era, upwards of four thousand have approached the Sun within the orbit of Mars. Of this number, observations of 607 have been made in Europe and China up to the middle of the present century. The greater number of them have been visible to the naked eye, the telescope never having been much used in searching for them before the middle of the last century. Small telescopic comets are very common, and a year scarcely ever passes away without two or three, and sometimes more, being discovered. In short, although the general public are not aware of the fact, astronomers have usually one or more of these faint bodies on their list of observable objects.

The head, or nucleus, of a large comet has mostly a central condensation of light surrounded by a nebulosity more or less luminous. When viewed through a low-power telescope this nucleus appears tolerably well-defined, with a faint nebulous tail, generally of several degrees in length, streaming into space in a direction opposite to the Sun. When, however, a powerful telescope is employed with a higher magnifying power, the nuclei of no two comets are similar, and even that of the same comet will change from day to day, and sometimes from hour to hour. Changes have even been noticed while the nucleus has been under the eye of the observer.

Comets perform their journey around the Sun in very eccentric orbits inclined to the ecliptic in all directions. The time occupied in their revolution is very different. The greater number have what is called parabolic orbits, which show that their motion during the time they have been under observation forms a part of a parabola. In these cases the period of revolution may extend over hundreds, or thousands, of years, and any prediction of their return is uncertain, but there is some probability of their reappearance at a distant period. Several have been observed to move in ellipses, and a few in hyperbolas. Comets which move in ellipses must return periodically at epochs more or less distant. Most of the brilliant comets which have been remarkable have parabolic orbits, that predicted by Halley to appear in 1759 being the only exception. The periodical comets are generally telescopic; so faint are they that it requires the greatest skill of the observer to detect them among the stars, and when found to fix their positions. They are, nevertheless, much more interesting to the astronomer than some of the grandest. That known by the name of Encke's comet is peculiarly so. This object was observed on several occasions between 1786 and 1818 with the belief that in each year a different comet had been observed. It was not till after the discussion by M. Encke of the observations made in 1818, that it was proved by that astronomer that this was a comet of short period, and that it revolved around the Sun in an eccentric ellipse in little more than three years. In 1822, 1825,

A, THE COMET OF 1861. B, NUCLEUS OF COMET OF 1861. C AND D, THE COMET OF 1858. E, JUPITER.

FF
281

1828, 1832, and at regular intervals up to the present time this little filmy object has returned to its perihelion, or nearest distance to the Sun. At these times it has always been eagerly sought for, and its orbit has been computed, so that its motion in space is now known with nearly the same accuracy as that of an ordinary minor planet. There are several other comets of short period whose orbits are tolerably well known, as those discovered originally by MM. Faye, De Vico, Brorsen, Winnecke, and Biela, all of which are faint telescopic objects. That of Biela is remarkable for having divided itself into two comets, since which it appears to have been entirely dissolved. This comet had faithfully returned, according to prediction, on several occasions, but in 1865, when its reappearance was expected, no trace of it could be seen even in the most powerful telescopes. Halley's comet, whose period is nearly seventy-seven years, has been so well observed, both in 1758-9 and 1835-6, that its return in 1912 may be looked upon as a matter of certainty.

Although the periodical comets are so peculiarly interesting to the astronomical observer, yet it cannot be denied that the popular interest rests entirely in favour of that class which has a more sensational appearance. The astronomer also takes this opportunity to observe the varying forms of the nucleus, and to determine, if possible, the nature of its physical composition. We have remarked in a preceding chapter that the spectroscope has enabled Mr. Huggins to observe that the spectrum of a small comet, discovered in 1868, is coincident with that produced by the combustion of carbon. How such a material should compose the head of a comet is difficult of explanation, and must remain for the present a matter for future research. On the next appearance of a brilliant comet, it is expected that the analysis of its light by means of the spectroscope will add considerably to our present imperfect knowledge of the constitution of these bodies, which have thus been proved to shine by their own light, and not by the reflected light of the Sun.

The most remarkable comets which have appeared during the present century are those of 1811, 1843, 1858, and 1861. That of 1811 was situated in a very favourable position for observation, having been always above the horizon of Great Britain during several months. It was first detected by M. Flauguergues on the 26th of March, 1811, in the south of France, and it remained visible until August, 1812. It was at its greatest brilliancy about October, 1811. At this time the length of the train of nebulous light was upwards of one hundred millions of miles, and its breadth about fifteen millions. It was remarked by M. Burckhardt that the tail was not immediately connected with the comet, but that it formed, at some little distance from the nucleus, a wide belt, which girded the nucleus in the same manner as the rings of Saturn. The colour of the head appeared to be greenish, or bluish green, giving rather a curious effect. Sir William Herschel noticed that there seemed to be an accumulation of matter towards the Sun in the head of the comet, the diameter of which was about one hundred and twenty-seven thousand miles. By viewing the comet with a night-glass of low power, Sir W. Herschel found the tail was generally inclosed at the sides by two condensed streams of light. Being very near the Milky Way, a comparison of the comet with a portion of that luminous stratum where no stars can be seen by the naked eye, showed that the two were of equal relative brightness.

The comet of 1811 was supposed to have had a considerable influence on the weather of that year, and upon the bountiful harvest and vintage, which appear to have been almost universal. Even now we often hear a reference made to the celebrated comet year, and wines have been sold at unusually high prices on account of some favourable influence produced by this popular comet. Other comets of later years have, however, been reputed to have had similar influences on the weather and the fruitfulness of agricultural produce. It is not always agreeable to express any doubt of the truth of some of the most popular beliefs in the supposed influences of the heavenly bodies on the atmosphere of the Earth. But it is nevertheless a fact, determined from several long series of observations, that neither the Moon, planets, stars, nor comets, give out sufficient heat to affect the most delicate mercurial thermometer yet constructed, and that any supposed influence which they may individually have upon the weather is only a popular fallacy resting on no real foundation. It is a well-known meteorological fact that the summer of 1811 was more than usually hot; but this coincidence with the appearance of a great comet can only be assumed to be accidental. The following lines by Hogg, the Ettrick Shepherd, refer to the appearance of the magnificent comet of 1811:—

> "Stranger of heaven, I bid thee hail ;
> Shred from the pall of glory riven,
> That flashest in celestial gale,
> Broad pennon of the King of heaven.
>
> "Whate'er portends thy front of fire,
> And streaming locks so lovely pale ;
> Or peace to man, or judgments dire,
> Stranger of heaven, I bid thee hail."

Those who witnessed the great comet of 1843 generally agree that it was the finest seen in the present century. Its full brilliancy was not visible in northern latitudes, where its luminous train first attracted attention while the nucleus was below the horizon. The tail was at that time, March 17th, about forty degrees in length, stretching from the horizon over the south-western sky as far as the constellation Lepus. Great excitement was caused by the sudden apparition of this long train of nebulous light, which at first was taken by some observers as belonging to the zodiacal light, which is generally most visible in the west soon after sunset at this time of the year. Throughout the southern hemisphere this comet presented a splendid appearance in the first days of March. Although the nucleus was not of great magnitude, yet it was extremely bright, and distinctly coloured; according to some observers, of a golden hue, similar to that of Venus; and according to others it was tinged with red. The disk had a well-defined planetary appearance, and was estimated to be about 4,500 miles in diameter, or rather more than half the size of the Earth. Sir Thomas Maclear, Government Astronomer at the Cape of Good Hope, observes that "of the casual observatory phenomena, the grand comet of March takes precedence; and few of its kind have been so splendid and imposing. I remember that of 1811; it was not half so brilliant as the late one. Immersed in the ravines of the Cedarberg, with high and precipitous ranges on each side of me, I made strenuous efforts to reach the Snewberg station, to command a view of the sudden visitor. Those unacquainted with the character of the Cedarberg cannot form a conception of the difficulties I had to encounter. For seventeen days we toiled on, tantalised every evening by seeing a portion of the tail over the mountain tops, and sometimes a sight of its bright head, as openings in the mountains permitted." The intrinsic lustre of the nucleus was so great, that the comet was seen in full daylight, when at a short distance from the Sun, by several observers in different parts of the world. Mr. Hind remarks that, "according to the most trustworthy calculations, the perihelion distance was only 538,500 miles; therefore the centre of the comet would be distant from the *surface* of the Sun's globe less than 96,000 miles!"

The great comet of 1858, known as Donati's comet, was visible to the naked eye in northern latitudes during at least six weeks, and, as a telescopic object, much longer. It was first seen by M. Donati, of Florence, on the 2nd of June, when it was very faint. For some time it gave no indication of becoming the splendid object which it presented from the middle of September till the second week in October. On the 23rd of September, the Moon being nearly full, the head of the comet appeared about equal to a star of the first magnitude, and was clearly visible half an hour after sunset. The greatest length of the train was observed on October 10th, when the main stream of light could be distinguished through an arc of sixty degrees, equal to about fifty millions of miles, the breadth of the widest part being about ten millions. Although the comet of 1858 has been surpassed by others in magnitude, yet the intensity of its nucleus will bear comparison with any previously observed comet. The nebulosity surrounding the head of the comet of 1811 was considerably larger; but including all the attractive features exhibited by Donati's comet, the popular interest evinced by the appearance of the two appears to have been about equal. It was fortunate that during the time when the comet of 1858 was near the star Arcturus, which was enveloped in the tail on October 5th, the Moon was absent from our evening skies. The effect of moonlight would have been to cut off at least two-thirds of the train, and a large proportion of the brightness of the remaining third. The tail was subject to constant changes of form. It was sensibly curved, and it had the appearance of an ostrich feather being gently carried through the air by the hand. At one time the train was bifurcated, issuing from the head in two unequal streams, forming its two sides, a dark space being left between them behind the nucleus. This dark space is delineated in most of the later telescopic drawings of the comet. On the 5th of October, when at its greatest brilliancy, the extremity of the train reached the two southernmost stars in the tail of Ursa Major, the head being in Boötes a few degrees south of Arcturus. Between the 16th and 20th of October, the comet ceased to be observed in Europe and North America. It was first seen in the southern hemisphere in the second week of October, and telescopic observations were continued at the Cape of Good Hope till the 4th of March, 1859.

The telescopic appearance of Donati's comet was remarkable in consequence of the continual changes which the nucleus and envelope were undergoing. As seen in one of the telescopes at the Royal Observatory on October 2nd, at 7 p.m., the nucleus presented a bright stellar point surrounded by an annulus of light, about four times its diameter, considerably fainter than the nucleus, but brighter than the general envelope, and about two-thirds complete. The Astronomer Royal divided the head into the following parts:—"1st. The parabolic envelope and inclosed illumination. 2nd. A brighter flat circular disk laid upon the inclosed illumination, just touching the parabola at its vertex. 3rd. A still brighter, flat, circular disk, concentric with the last, about one-fourth of its diameter; no bright ring. 4th. The nucleus, concentric with the last, about one-third of its diameter, well-defined, and looking very hard. From the nucleus a dark shadow diverged, cutting off the light of the circular disks, and everything except the nucleus itself." On October 3rd, several of these appearances had considerably changed, the matter composing the nucleus and envelope being evidently in a continual state of local excitement.

The orbit of this comet has been most accurately computed from the numerous observations made in Europe and North America. It seems to show that this is a periodical comet, but how far it goes into space in the interval during its absence from us no one can tell. It is, however, expected that in two thousand years hence, if nothing unforeseen happens to it, it will return once more to its perihelion passage around the Sun, when it will again adorn our skies. Meanwhile, many years will probably elapse before the remembrance of this beautiful celestial visitor will be erased from the memories of all those of the present generation who were so fortunate as to witness it.

The comet of 1861, although less attractive than the three just described, was a very interesting object. It was discovered in Europe low down in the north on the evening of June 30th, by several people. M. Secchi first took the immense train of light for the smoke of fireworks, which they often have at Rome on the festival of St. Peter and St. Paul; but he soon perceived that the light continued, when he judged that it must be a comet, although the head was too near the horizon to be well observed. About midnight on the same day it was seen at the Royal Observatory, Greenwich. The telescopic appearance was generally observed to be subject to similar changes to those noticed in former comets; the nucleus was, however, somewhat elliptical, its light nearly equalling that of Saturn. It was well defined on all sides, but on the side nearest the Sun it was less sharp, jets of reddish light starting from its edge in the form of a fan. It is a remarkable fact that on the morning of June 29th, the Earth was very near the extremity of the tail; and Mr. Hind has suggested that a peculiar illumination of the sky which he noticed about this time was possibly "attributable to the commingling of the matter forming the tail of the comet with the Earth's atmosphere."

Although the great comet of 1861 was not visible in Europe till the evening of the last day in June, yet a young amateur astronomer, Mr. Tebbutt, of Windsor, New South Wales, saw it so early as May 13th. The telescopic appearance of this comet was remarkable for the numerous jets of light thrown out from the nucleus. An example of this is exhibited in the woodcut on page 281. These jets of light were very distinct, and often varied in form in short intervals of time, even while the eye of the observer was directed to them. The Rev. T. W. Webb, referring to the different envelopes which were exterior to the nucleus and jets of light, remarks: — "The number and complex arrangement of these luminous veils on June 30th, as seen in the comet eye-piece, with a field of about 52′ and a power of 27, produced a singular as well as magnificent effect; it was as though a number of light hazy clouds were floating around a miniature full Moon. Portions of six could be more or less distinctly traced. The innermost of these was very narrow and short, faint on the left hand, but brighter on the right, in the inverted view, and coming up close to the nucleus. The second was a parabolic arc in which the nucleus stood." On the next night of observation, July 2nd, the envelopes had considerably changed, but they still retained their general form.

From a comparison of some drawings of the comets of 1858 and 1861 made at the Royal Observatory, the comet of 1858 is the more attractive of the two with regard to their general appearance; but that of 1861 is the more interesting as a telescopic object, so far as relates to the continual and fantastic changes in the form of the nucleus and the numerous jets emanating from it.

METEORS

or

SHOOTING STARS.

Who forthwith from the glittering staff unfurled
The imperial ensign; which, full high advanced,
Shone like a meteor streaming to the wind,
With gems and golden lustre rich imblazed
Seraphic arms and trophies."

Paradise Lost.

PERIODICAL METEORS.

———◆———

OPULAR attention has been directed to these minute cosmical bodies during the last few years, owing to the appearance, according to prediction, of the memorable star-shower which interested so many on the night of November 13th, 1866. The thousands of luminous particles of matter which were finally destroyed at that time, by coming in contact with the Earth's atmosphere, seemed to have made no impression on their numbers, for on the corresponding nights in 1867 and 1868 a repetition of the display was observed in various parts of North America, and in a less degree in other parts of the globe. There is no doubt whatever that the memorable stream of meteors which crosses the Earth's path in November, consists of millions of these objects; and it is possible that during the progress of the principal part of the stream in its course around the Sun, a far greater number are formed than those extinguished by combustion in the Earth's atmosphere.

One of the first who drew particular attention to the great display of meteors which was expected to take place in November, 1866, was Professor H. A. Newton, of Yale College, Newhaven, U.S. This gentleman contributed, in the year 1864, two remarkable papers on "November Star-showers," in which he gave a determination of the length of the meteoric cycle, the annual period, and the probable orbit around the Sun. Though his final results have now been considerably modified by subsequent observations, yet to Professor Newton belongs the honour of being the first person who clearly demonstrated the periodical nature of these minute bodies, by the publication of approximate elements of the supposed orbit of the November stream. Some idea may be gathered of the difficulty of the problem undertaken and developed by Professor Newton, and afterwards by other mathematicians, if we quote the following words of Humboldt:—"Although the asteroid swarms approximate to a certain degree, in their inconsiderable mass and the diversity of their orbits, to comets, they present this essential difference from the latter bodies, that our knowledge of their existence is almost entirely limited to the moment of their destruction; that is, to the period when, drawn within the sphere of the Earth's attraction, they become luminous and ignite."

It will probably be interesting if we give brief notices of a few of the principal authentic November star-showers. It is from these data that the first approximation of the period of revolution of this cosmical ring has been obtained. Most writers on this subject have already alluded to these phenomena recorded here and there in old histories. We do not, however, hesitate to reproduce one or two of them once more. The years in which recorded November star-showers took place, are 902, 931, 934, 1002, 1101, 1202, 1366, 1533, 1602, 1698, 1799, 1832, 1833, 1866, 1867, and 1868. Condé, in his *History of the Dominion of the Arabs in Spain*, mentions that on the death of King Ibrahim bin Ahmad, in the year 902, about the middle of

METEORS NEAR THE RADIANT POINT IN LEO, AS SEEN FROM THE ROYAL OBSERVATORY, GREENWICH.

October, o.s., "an infinite number of stars were seen during the night, scattering themselves like rain to the right and left, and that year was known as the year of the stars." An Arab writer relates of the same phenomenon that "in this year there happened in Egypt an earthquake, lasting from the middle of the night until morning; and so-called flaming stars struck one against another violently while being borne eastward and westward, northward and southward; and none could bear to look towards the heavens." These accounts from Mohammedan sources, though bearing evident marks of superstition and exaggerated observation, are sufficient to show that a probably unusual meteoric display took place in the year indicated.

We have another Mohammedan record that "in the year 599 (A.D. 1202), on the night of Saturday, on the last day of Muharram, stars shot hither and thither in the heavens, eastward and westward, and flew against one another, like a scattering swarm of locusts, to the right and left. This phenomenon lasted until daybreak. People were thrown into consternation, and cried to God the Most High with confused clamour; the like of it never happened except in the year of the mission of the Prophet, and in the year 241."

Any unusual celestial phenomenon, whether it be a comet, meteor, or eclipse of the Sun or Moon, was always considered in former ages as some omen portending good or evil to the nation. This superstitious feeling is clearly shown in an extract from a Portuguese work, *Chronicas dos Reis de Portugal Reformadas*, 1660, relating to an unusual star-shower, observed nearly three hundred years previously. "In the year 1366, and xxii days of the month of October (o.s.) being past, there was in the heavens a movement of stars, such as men never before saw or heard of. From midnight onwards, all the stars moved from the east to the west; and, after being together, they began to move, some in one direction, and others in another. And afterwards they fell from the sky in such numbers, and so thickly together, that as they descended low in the air they seemed large and fiery, and the sky and the air seemed to be in flames, and even the Earth appeared ready to take fire. That portion of the sky where there were no stars, seemed to be divided into many parts, and this lasted for a long time. Those that saw it were filled with such great fear and dismay that they were astounded, imagining they were all dead men, and that the end of the world had come." This apparently improbable account of an extraordinary star-shower may be sensibly correct if we assume that this great meteoric display was accompanied by a coloured aurora, the streamers of which, with very little imagination, might be suggestive of fire.

It is not necessary to notice any more of these ancient meteoric showers, excepting only to state that the dates have been satisfactorily proved, and that they have done good service as data for modern calculations.

Of modern displays, Humboldt's description of that of November 12th, 1799, as seen by him and M. Bonpland at Cumana, South America, is well known. During four hours thousands of falling stars were noticed. M. Bonpland states that there was not a space in the firmament equal in extent to three diameters of the Moon, that was not filled at every instant with bolides and falling stars. The meteors were visible till sunrise. The same star-shower was observed at other places in South America, some of them distant 700 miles from Cumana.

Our illustration gives an idea of this extraordinary display as witnessed by Mr.

Andrew Ellicott, at sea, in lat. 25° N., off the coast of Florida. He has recorded his observations in his journal as follows:—" November 12th, 1799, about 3 a.m., I was called up to see the shooting of the stars, as it is commonly called. The phenomenon was grand and awful; the whole heaven appeared as if illuminated by sky-rockets, which disappeared only by the light of the sun after daybreak. The meteors, which at any one instant of time appeared as numerous as the stars, flew in all possible directions, except from the Earth, towards which they all inclined more or less."

This shower was visible partially in England and in Greenland. In November, 1833, Mr. Olmsted and another observer are said to have observed at least 240,000 meteors in a period of nine hours. The wonderful exhibition of falling stars in the year 1833 was quite, if not more remarkable, than that of 1799. Major Strickland, in a work published twenty years afterwards, entitled *Twenty-seven Years in Canada West*, gives a most graphic account of this memorable star-shower as follows:—

"I think it was on the 14th* of November, 1833, that I witnessed one of the most splendid spectacles in the world. My wife awoke me between two and three o'clock in the morning to tell me that it lightened incessantly. I immediately arose and looked out of the window, when I was perfectly dazzled by a brilliant display of falling stars. As this extraordinary phenomenon did not disappear, we dressed ourselves and went to the door, where we continued to watch the beautiful shower of fire till after daylight. These luminous bodies became visible in the zenith, taking the north-east in their descent. Few of them appeared to be of lesser size than a star of the first magnitude; very many among them seemed larger than Venus. Two of them in particular appeared half as large as the Moon. I should think, without exaggeration, that several hundreds of these beautiful stars were visible at the same time, all falling in the same direction, and leaving in their wake a long stream of fire. This appearance continued without intermission from the time I got up until after sunrise. No description of mine can give an adequate idea of the magnificence of this scene, which I would not willingly have missed. This remarkable phenomenon occurred on a clear and frosty night, when the ground was covered with about an inch of snow."

It must be understood that the small cosmical bodies composing the November ring of meteors move in orbits, which cut the plane of the ecliptic near the point occupied by our Earth about the 13th or 14th of November, and, consequently, can only be seen yearly about these days. And it must also be borne in mind that the smaller showers of meteors which are frequently observed at other periods of the year belong to a collection of meteors also moving around the Sun in orbits, inclined, at different angles, to the ecliptic, each forming a system different from each other. In short, it has been found from observation that there are several of these rings or systems crossing the Earth's path at different periods of the year, but the November and August meteors are by far the most numerous. On these occasions, meteors

---

* The principal display took place on the morning of November 13th, 1833. This date is therefore one day in error.

METEORIC SHOWER, AS SEEN OFF CAPE FLORIDA.

become luminous owing to the great heat developed by friction produced by the enormous velocity with which they enter the Earth's atmosphere, where they burn for a second or two, and are then finally dissipated into smoke or dust.

On account of the apparent rapid motion of meteors, the only satisfactory observations made during great displays are those for the determination of the point in the heavens from which they radiate. This apparent orbital velocity is ascertained to be about thirty-eight miles a second, made up by the compound effect of the movements of the Earth and the meteors in their respective orbits, that of the Earth being direct, while that of the meteors is retrograde. But when, as in the case of the great periodical showers of November, the observer is enabled to multiply his observations almost to an indefinite extent, this great velocity does not materially interfere in obtaining a correct determination of the exact position of the radiant point. On the accuracy of our knowledge of the position of this point almost everything depends in the calculation of the dimensions and position in space of the meteoric orbit. It does not, however, require a practised astronomer to notice that, in such a display as that of November, 1866, the majority of the meteors emanated from one spot in the heavens, for that was soon evident to the senses of the most casual observer. It is found, from the average of the observations made in that year by several astronomers, that the exact position of this radiant point of the November meteors is 149° 12′ of right ascension, and 23° 1′ of north declination. Any person having access to a celestial globe or map, can identify for himself this position in a moment. He will see that it is situated north of Regulus, between the well-known stars Gamma and Epsilon Leonis. From this accurately-observed radiant point of the November meteors, much of our present knowledge of the movement of these bodies is derived. It may be mentioned here that the radiant point of the August meteors is situated in the constellation Perseus. Of meteors in general, about sixty radiant points have been determined with more or less certainty: some of these have, however, been inferred from isolated observations of meteors noticed on the same day in different years. In a great number of cases the general radiant points are nearly as well determined as in the older and better-observed showers of August and November.

As an illustration of the pains taken for the determination of the radiant point, we cannot do better than record the notes of Sir John Herschel: "During the superb display of meteors on the night of the 13th November, 1866, my attention was particularly directed to the determination of the exact situation in the heavens of the radiant point of their courses, which has been commonly stated to be coincident with that of the bright star Gamma Leonis. This, however, was certainly not the case; and I am enabled to say with perfect confidence that their courses diverged, with a very remarkable degree of agreement, from a point considerably higher in declination, and less advanced in right ascension. . . . Several circumstances enabled me to fix on the point with full assurance of its being the true radiant: First, the frequent out-shooting from a very near proximity to it, and once in several different directions, of a volley of meteors. Secondly, on two or three distinct occasions a meteor appeared in this very point, and in all these cases it was motionless, devoid of a train, and, on its extinction, left only a small nebulous light to mark the place of its appearance. Thirdly, on one occasion, a meteor shot forth horizontally at a distance of about six or eight degrees from the radiant, and disappeared after running a very short course, leaving a vaporous train which continued visible (fading gradually away) for $2^{m}\ 40^{s}$; this afforded time for tracing back the direction of the train (which did not alter), and its course, continued, passed through the spot in question."

The observations seem to show that the true line of direction in space of each meteor's flight lies in an opposite direction to the Earth's path, and that the meteoric orbit does not deviate to any great extent from a circular form. The absolute velocity of meteors is therefore nearly that of the Earth in its orbit, the direction of their revolution being retrograde. Their motion is thus independent of that of the Earth, meteors being in reality excessively small planets revolving with the same regularity as the larger planets around the Sun. This conclusion is confirmed by the fact that the two principal radiant points at the August and November epochs, are precisely those to which our globe is directed in its orbital revolution at these dates.

A question has been asked, that if there be a ring of these small cosmical bodies, how is it that we only pass through the stream once in thirty-three years? Now, although we use the word "ring," it must not be supposed that the meteors are equally distributed along it; on the contrary, it is known that the principal agglomeration of these bodies takes place only in a limited portion of this ring, and that comparatively few are situated in other portions of it. As the times of revolutions of the Earth and meteors are not identical, then, in one year, the Earth would pass through the principal portion of the stream; in the following year, through a part where the meteors are more scattered; in the third year, where the ring is thinner still; this progression going on year after year till the cycle of thirty-three years is completed, when the Earth will again pass through the principal portion of the stream, and an unusual display will take place.

With regard to the number of meteors counted during the great shower of November, 1866, much depended upon the number of persons making the observations. At the Royal Observatory, Greenwich, nearly nine thousand were counted; and it is supposed that at least another thousand may have escaped the attention of the eight observers employed. At Edinburgh more than eight thousand were registered. At other stations the numbers reported varied only with the strength of the observing staff. Between 1^{h} a.m. and $1^{h}\ 15^{m}$ a.m. the principal part of the display took place. The average numbers observed per minute at Greenwich during the display can be clearly understood by reference to the accompanying diagram.

The star-shower of November, 1866, was confined generally to the eastern side of the Atlantic, while those of November, 1867 and 1868, were seen to the best advantage in the Northern States of America. Of the numerous accounts published in the public journals of the shower of 1866, the following extracts from those of the Rev. R. Main, Mr. Hind, Mr. A. S. Herschel, and Professor C. P. Smyth, will be sufficient examples of what was seen by different eyes on this memorable night. The Rev. R. Main, Radcliffe Observer at Oxford, remarks that "this great display began about 13ʰ (or 1 o'clock in

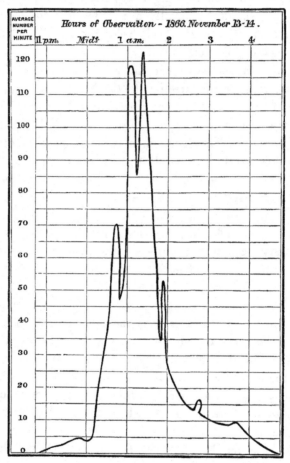

Diagram showing the average number of meteors per minute observed at the Royal Observatory, Greenwich, on the night of November 13—14, 1866.

the morning) and reached its maximum at about 13ʰ 24ᵐ, after which time it gradually began to slacken. The watch, however, was kept up till 18ʰ, though after 15ʰ there were not many meteors seen. In all, there were observed not fewer than 3,090 during the night, of which about 2,000 fell between 13ʰ and 14ʰ, or between 1 a.m. and 2 a.m. As to the general appearance of the meteors, it was noticed that the majority of them were of a whitish or yellowish colour. Some, however, were reddish or orange-coloured, and one meteor was noticed to be bluish. The brightest left generally a train behind them, which was to be seen for a few seconds after the meteor had disappeared.

"In one particular instance, the train of a meteor was visible for some minutes; this was a very bright meteor which disappeared in the belt of Orion, leaving the train apparently attached to Zeta Orionis, and giving to that star the appearance of a comet with a tail of nearly 3°, standing out at a position angle of nearly 135°. It then detached itself from the star, keeping up the same route as the meteor, but forming itself into a ball of faint cometic appearance of about 15′ diameter, which grew dimmer and more diffused, and disappeared altogether after a lapse of about 4^m to 5^m at a distance of nearly 1° from Zeta Orionis, at a position angle of about 200°. This meteor appeared at the time of the greatest display, at 13^h 24^m. Only in two instances meteors were seen to burst, one in the east and another in the north."

Mr. Hind made his observations at Twickenham, assisted by three observers, including M. Du Chaillu, the African traveller. From the statement of Mr. Hind we find that "from midnight to 1 o'clock a.m., Greenwich time, 1,120 meteors were noted, the number gradually increasing. From 1 a.m. to 1^h 7^m 5^s no less that 514 were counted, and we were conscious of having missed very many, owing to the rapidity of their succession. At the latter moment there was a rather sudden increase, to an extent which rendered it impossible to count the number, but after 1^h 20^m a decline became perceptible. The *maximum* was judged to have taken place at about 1^h 10^m, and at this time the appearance of the whole heavens was very beautiful, not to say magnificent. Beyond their immense number, however, the meteors were not particularly remarkable, either as regards brilliancy, or the persistence of the trains, few of which were visible more than three seconds: indeed, M. Du Chaillu observed that in these respects the meteors fell far short of those of the April period, which he had witnessed under a fine sky in equatorial Africa. From 1^h 52^m to 2^h 9^m, 300 were registered; from 3^h 9^m to 3^h 24^m, 100; from 4^h 42^m to 5^h, the number seen was 12, and these mostly faint; and from 5^h 45^m to 6^h, only five were counted.

"No persons acquainted with the constellations, who carefully watched the display, could have any doubt as to the accuracy of the astronomical theory relative to these bodies. The radiant in Leo was most strikingly manifested; while the meteors in the opposite quarter of the sky traversed arcs of many degrees, in the vicinity of the diverging point they shone out for a few seconds without appreciable motion, and might have been momentarily mistaken for stars.

"Several very vivid flashes of lightning were remarked during the night. The last, at 3^h 54^m, was particularly brilliant, of a deep orange colour, and apparently emanated below the radiant in Leo. The horizon in that quarter was occupied by a pale glow, resembling what has often been remarked during the aurora borealis."

Mr. Herschel observed that "the majority of the meteors seen were white, but some were of an orange colour. A large proportion of the meteors were brighter than the fixed stars. Four of them were much brighter than the planets. Their vivid streaks of greenish light, lasting on an average three seconds, but occasionally

H H

much longer, and extending from 20° to 40° in length, marked the tracks of many meteors, whose flight would otherwise have passed unnoted. From six to ten streaks could in this manner be counted together in the sky at times when the shower was most intense. At $1^h 15^m$ a flight of six or eight meteors, for the most part brighter than the fixed stars, started together from near the constellation Leo, which, by their orange-coloured light, and by the strong emerald green streaks which they left behind them, gave the impression of a burst of artificial rockets.

"Large meteors appeared at $12^h 33^m$, $12^h 41^m$, $2^h 14^m$, and $2^h 42^m$. These meteors left streaks which endured from five to fourteen minutes. Each of the eccentric streaks contracted itself into a knot of light, which drifted, cloud-like, eight or ten degrees before it vanished. The last presented an imposing ring of flame, three or four degrees in width. It remained visible for fourteen minutes, gradually expanding itself into a heart-shaped loop, until it enclosed the chief stars of Ursa Major in a wide fantastic wreath before it disappeared.

"The meteors radiated from the usual point of divergence of the shower, and shot to all parts of the sky with a swift and stately motion."

The numbers observed by Mr. C. P. Smyth and those associated with him, were 8,312. He remarks that "the general characters of these meteors were bright yellow balls, like Jupiter or Venus for brightness, but attended by long trains of faint and light-blue light; these trains usually lasted only from two to three seconds, were of an exaggerated long elliptical form, so as to be nearly invisible close to the bright head, and to be broadest near the middle of their length; and such broader central part of the train often remained abundantly visible, long after the head and chief part of the length of the train had entirely disappeared: so decidedly too was this a material fact, and not any optical impression caused by the overpowering brightness of the head during the time it was visible, that often, on turning to a new part of the sky, the last expiring traces of such short central part of a meteor track were seen, proving that a meteor had just passed that way, and had been missed by the observer. . . . . . . Without knowing anything at all of the actual distances of these bodies, they gave the impression of some ethereal description of rockets, but endued with the speed of cannon-balls, and half their purity of path; and one of them often followed another, second after second, for several seconds together, as if discharged from the other side of the sky at something positive in the west, and with a determination to hit it too. One, and only one, case of directly retrograde motion was observed."

The star-shower of November, 1866, was observed in all parts of Europe, and even to the southernmost portion of Africa. At the Cape of Good Hope, Mr. G. H. Maclear and one observer counted 2,742 meteors. Some were of an orange colour with long greenish trains; some a deep red; others without any train at all; forming, as Mr. Maclear remarks, "one of the most magnificent spectacles that it has been the fortune of man to witness." In November, 1867 and 1868, the number of meteors counted by observers in North America amounted to several thousands.

A word or two may be said on the observed heights of meteors. During great displays, this is not very easy to ascertain, on account of the immense number of observations making individual identification a difficult matter. When meteors are fewer, many simultaneous observations have been made; for it is necessary that the same meteor be carefully observed in several localities, separated from each other by a considerable distance. In November, 1865, fifteen meteors were identified as having been observed at Hawkhurst, by Mr. Alexander Herschel, and at Cambridge by Professor Adams. The average height of these meteors was about seventy miles. On August 10th, 1863, some corresponding observations of this kind were made at Greenwich, Cambridge, and on the south coast, from which Mr. Herschel deduced an average height of eighty miles. These meteors attracted the attention of the observers on account of their extreme brilliancy. One of these was observed at Cranford, Middlesex, and at Hawkhurst with great precision. The beginning and end of its apparition as well as its direction among the stars were recorded at both places, the resulting height being seventy-nine and fifty-eight miles respectively. The velocity was observed to be thirty-eight miles a second. A very extraordinary single meteor was seen in many parts of England on February 11th, 1850, and created considerable interest. It exploded near Biggleswade, about nineteen miles only above the Earth's surface.

Humboldt made some careful observations respecting the colour of meteors. He found two-thirds were white, the remainder yellow, yellowish red, and green. At Greenwich, the majority have been registered of a blue colour; the remainder white, bluish white, yellow, red, and green. On one or two occasions during the display of 1866, meteors were observed to have a double outburst, the principal meteors passing on at the head of the stream of light, the secondary nucleus remaining in the luminous track.

Numerous analyses of meteors, portions of which have fallen to the ground, have been made by chemists, both English and foreign. A few of these are referred to in our remarks on aërolites and bolides. It is supposed that the material forming the thousands of shooting stars visible on extraordinary occasions is of the same nature as that composing an ordinary aërolite. If so, their chemical elements consist of similar matter to that distributed throughout the Earth's crust, for analysis has shown that they contain indications of the presence of iron, copper, zinc, nickel, sulphur, cobalt, arsenic, manganese, chromium, potash, phosphorus, soda, and carbon.

We now come to a most important speculation, first published by M. Schiaparelli, of Milan, with reference to the possible connection of meteors with comets. In other words, is it possible that a comet is composed of an accumulation of meteors, or that one large meteor is acting as a nucleus to a system of these minute bodies? M. Schiaparelli's researches certainly give strong evidence that this question ought to be answered in the affirmative. The Italian astronomer may be, and probably is, right in his deductions, for it is generally acknowledged by those competent to give an opinion on such a speculative subject, that his discovery of the coincidences of the orbits of the streams of meteors with those of certain observed comets must be classed among the most remarkable in modern astronomical researches. From subsequent investigations of M. Le Verrier and Professor Adams, the deductions of M. Schiaparelli appear to be so far borne out, that we think the subject may even almost be removed from the section of "speculative" astronomy. In consideration of the great importance of M. Schiaparelli's researches, we will endeavour to give a brief explanation of what has been done on the subject.

First, may we assume it impossible to imagine the existence of a system of minute bodies grouped around one or more nuclei, as, for instance, a body of meteors around one or more comets? for between the systems of comets and meteors many analogies have been recognised for many years past. M. Schiaparelli considers that if a mixed system of this kind were brought near our Earth by the attraction of the Sun, the orbits described by its principal members would differ but little from those of the smaller bodies. It is evident, if this reasoning be true, that when we find the elements of a meteoric ring identical in magnitude and position with those of any known comet, we may, without much stretch of imagination, infer that the comet forms part of that ring, and would be one of its constituent members.

In the first instance, M. Schiaparelli selected for examination the August ring of meteors, whose radiant point is in the constellation Perseus. He then calculated the elements of its orbit. On comparing these elements with those belonging to a large comet, which was visible to the naked eye in 1862, they were found to be almost identical. In short, this comet and the meteors appeared to be revolving around the Sun at the same average distance, and in orbits of the same form, and inclined to an equal amount to the ecliptic. To show this wonderful agreement, it will be necessary to give the elements in astronomical language. Our readers will see at once a great similarity in the corresponding parts of the two sets of elements.

	Ring of August Meteors.	Comet II., 1862.
Perihelion passage* . . .	1866, July 23·62 . .	1862, Aug. 22·9
Ascending node passage† .	Aug. 10·75 . .	. .
Long. of perihelion . . .	343°.28′ . .	344°.41′
Long. of ascending node .	138°.16′ . .	137°.27′
Inclination of orbit . . .	64°.3′ . .	66°.25
Perihelion distance . . .	0·9643 . .	0·9626
Direction of motion . . .	Retrograde . .	Retrograde

It is probable that the differences in these two sets of elements are no more than accidental. M. Schiaparelli finds that if he makes a slight but reasonable alteration in his assumption, the differences will entirely disappear. He has therefore come to the

---

* *Perihelion* is that point in the orbit at which a planet or comet is nearest to the Sun.
† *Node* is the point where the centre of a planet or the nucleus of a comet crosses the ecliptic.

conclusion that the great comet of 1862 is no other than one of the August meteors, and is probably the most considerable in magnitude of the group.

The period of revolution of the comet, and also of the August meteors, is above one hundred years. M. Oppolzer has calculated that the minimum distance of this comet from the Earth's path is less than the lunar orbit. If this approaching the Earth were to take place on the 10th of August, shortly before noon, we should pass through a dense part of the comet.

Startling as this coincidence of elements appeared to be, and which was sufficient to draw the attention of some of our first mathematicians to this research, a second coincidence still more remarkable was announced, resulting from a discussion of the observations of the November star-shower of 1866. MM. Schiaparelli, Le Verrier, and Professor Adams, have each independently computed the elements of this great meteoric ring, all of which agree sensibly with each other. They are identical with those of a telescopic comet which appeared in January, 1866. There seems to be no doubt, therefore, that this comet and the November meteors are also revolving around the Sun in a common orbit. The following are the elements of each:—

	Ring of Nov. Meteors.	Comet I., 1866.
Perihelion passage	1866, Nov. 10·09	1866, Jan. 11·16
Period	33·25 years	. 33·18 years
Mean dis. from Sun (Earth's = 1) . .	10·340	. 10·325
Eccentricity	0·904	. 0·905
Perihelion distance	0·989	. 0·977
Inclination of orbit	14°.41′	. 17°.18′
Long. of ascending node	51°.18′	. 51°.26′
Long. of perihelion	42°.24′	. 42°.24′
Direction of motion	Retrograde	. Retrograde

A mere reference to these figures is sufficient to see their great similarity.

It has now been decided that the periodic time of the November star-shower is not 354·6 days, which was the interval assigned originally by Professor Newton. The true period is about thirty-three and a quarter years. Professor Adams has made some important calculations on the effects produced by the action of Jupiter, Saturn, and Uranus on this ring, while the principal stream passes through space in the neighbourhood of these large planets. He has conclusively shown that, during a period of thirty-three and a quarter years, the longitude of the node is increased 20′ by the action of Jupiter, nearly 7′ by the action of Saturn, and about 1′ by that of Uranus. The other planets produce scarcely any sensible perturbations; so that the computed increase in the longitude to be taken into account through these planetary disturbances amounts to about 28′. The observed increase of longitude in the same interval of time is 29′. This remarkable agreement between observation and theory enabled Professor Adams to announce his belief in the absolute correctness of the period of thirty-three and a quarter years.

We have thus briefly shown that the remarkable coincidences in the elements of certain comets and meteors, originating in the speculative mind of M. Schiaparelli, have occupied the attention of our most eminent mathematicians. The general result of their investigations is to confirm those of the Italian astronomer. But time must settle whether the connection of comets with meteors is as real as these recent observations and researches would lead us to suppose. If the similarity of orbits should, after all, turn out to be merely accidental, we must remark that it is one of the most extraordinary of all astronomical coincidences; for it seems scarcely possible, in the doctrine of chances, that the computed elements of the orbits of several comets should be so exact a counterpart of those of corresponding streams of meteors.

Now that the August and November streams have each their comet, may we not infer from analogy that the remaining systems are similarly accompanied? A known comet is suspected to belong to the April ring, but the observations are too few to settle the point. M. Schiaparelli remarks, "I have shown the possibility that all these bodies, great and small, form in space systems drawn together by their own attraction, and afterwards dispersed by solar action. Perhaps, also, what we call a comet is not a single body, but an immense collection of small particles attached to a principal nucleus. Chladni, in his work *Die Feuermeteore*, has considered comets from this point of view. He has even propounded a theory of falling stars, differing but little from my own speculations. But Chladni wrote in 1819, when little was known of the theory of the movement of meteor-systems; by the skilful manner he handled the subject, he was evidently working with a mind beyond his age."

The November meteors will not in future be expected in any great numbers, until the principal part of the stream has performed another thirty-three years' journey around the Sun. The year 1900 is expected to be the date of the next grand display. Observations of meteors in other parts of this ring will, however, be observed as usual in small numbers at every succeeding November in the interval. The author has in former intermediate years recorded the positions of numerous meteors, evidently belonging to the thinner parts of this ring. At the Royal Observatory a careful watch is always kept up for the August and November meteors, and also occasionally when others are expected.

AËROLITES AND BOLIDES.

What is an aërolite? The general meaning of this natural phenomenon may be obtained at once from the derivation of the word, ἀήρ, the air, and λίθος, a stone. It may therefore be assumed that an aërolite was originally supposed to be a stone formed by some means in the Earth's atmosphere, from which it was precipitated to the ground. But since all meteors are now considered to be cosmical bodies moving independently in space, their appearance to us can better be explained by assuming that, as the meteor enters the higher regions of our atmosphere with an enormous velocity, the friction caused by its rapid motion soon ignites the material of which it is composed, when combustion and explosion is the result. All meteors, whether they be shooting-stars, aërolites, or bolides, belong to one common family, and are composed, for the most part, of similar substances. But an aërolite proper is a meteor which has accidentally approached within a few miles of the Earth, where an explosion having taken place, the fragments are scattered in various directions, some of which have been afterwards discovered. When these fragments have been found soon after their descent, they have generally felt warm, and sometimes even hot to the hand. On subjecting these aërolites to analysis, they are always found to be composed, either of a rocky substance or of iron ore; but the relative proportions of these elements in different specimens are very variable. Occasionally they are formed principally of earthy or rocky substances, with little iron; at other times the iron-stone preponderates, making the aërolite simply a metallic mass; while in the majority of cases the metallic and stony ingredients are indiscriminately mingled together in about equal proportions.

In the latter part of the eighteenth century, so little trustworthy evidence had been collected in relation to meteors that scarcely any person had a true conception of the origin of aërolites, and all accounts of their fall to the earth were treated as tales originating from ignorance and superstition. At the present time, most of these old accounts are believed to be generally authentic; at all events, nothing recorded in them is more improbable than in many modern falls, the accuracy of which have been proved beyond dispute. At the first announcement of the celebrated fall of stones at L'Aigle, Normandy, in 1803, the subject afforded much amusement to the wits of Paris, and compassion was expressed in the newspapers, that a mayor of so important a place should be so credulous as to believe such absurd nonsense as that stones had fallen from the clouds. There was, however, one philosopher who published a tract, nine years previously, on a large metallic mass said to have fallen from the sky in Siberia, in which he announced his belief in the possibility of the accuracy of the traditions held by the Tartars respecting it.

It is to Chladni, therefore, to whom we owe the first reasonable hypothesis of the origin of aërolites, which, although it failed at first to make much impression on the minds of scientific men, drew eventually the attention of philosophers to the subject. In 1802, Edward Howard, F.R.S., read a paper before the Royal Society, giving an account of the analysis of an aërolite which fell near Benares in 1798. The result of his experiments appeared to give so satisfactory a conclusion as to the non-terrestrial origin of this specimen, that men of science generally gave in their adhesion to his opinion. When the memoir of Mr. Howard appeared in Paris, the French chemists were disposed at first to disagree with the opinions expressed by the English analyst, although his result showed that the meteoric substances bore but little resemblance to any known minerals. A remark of M. De Laplace, however, convinced the French *savans* of the possibility of these stones not belonging to our globe. "It is possible," he says "for stones to be thrown upon our Earth by volcanoes in the Moon. Do not reject, therefore, as impossible a fact which deserves to be carefully examined; gather together all the facts of this kind, endeavour to discover the truth, and if terrestrial physics cannot explain to us the origin, we shall be able to find it in celestial physics." After the fall of stones at L'Aigle, M. Biot was appointed by the Institute of France to examine on the spot all the circumstances relating to that remarkable shower of stones. The specimens collected by M. Biot were placed for analysis in the hands of MM. Vauquelin and Thénard, who found that the substances contained in them were very similar to those found by Howard in the Benares stone. Alluding to these investigations, the celebrated Cuvier, in a report on the progress of science in the ten years previous to 1809, states, "that the phenomenon of stones fallen from the atmosphere, known both in antiquity and during the middle ages, had only been established as truths in physical science, during that period, by the conjectures of Chladni, the analyses of Howard, Vauquelin, and Laugier, and the researches of Biot." The numerous authenticated falls of aërolites since the date of Cuvier's report, combined with the observed recurrence of extraordinary star-showers at regular periods of $33\frac{1}{4}$ years, have conclusively settled that these bodies exist and move in space, governed by the same laws as the planets of the solar system. The researches of Chladni, whose philosophic mind first saw the possibility of the existence of such a mass of minute planetary bodies, have been thus proved to have been based more upon inductive facts than upon the conjectural reasoning ascribed to him by Cuvier.

In Chladni's tract on "Masses of iron and stone reputed to have fallen from the air," he has expressed an opinion that all the accounts hitherto received of the falls of aërolites were correct. He also compiled a catalogue of them, including all the large fire-balls, accounts of which had been given in the works of various authors. It was a part of the hypothesis of Chladni that minute cosmical bodies exist throughout the solar system, all moving with a great velocity, and in orbits of widely different eccentricities around the Sun. He also considered that the vivid light and heat of combustion exhibited by these bodies when they are first brought into contact with the upper strata of the atmosphere are produced by a certain property of compressed air. Dr. Joule, of Manchester, made a series of accurate experiments, from which he has concluded that the great velocity of meteors through the atmosphere is sufficient to produce a heat upon the surface of the meteoric body of such an intensity as to fuse, and probably also to volatilise, the hardest substances. Not only do aërolites exhibit certain signs of having undergone the action of an intense heat, by the glazed surface or crust with which they are invariably covered, but the appearance of silent fire-balls, or bolides, and ordinary shooting-stars, can also be explained by similar assumptions.

Formerly, the records of a phenomenon of this nature were generally confined within a limited distance of the locality where it took place; but owing to a constant intercommunication between distant countries, these occurrences are now reported at once to the scientific world by means of the daily press. Falls of aërolites are much more common than they were supposed to be in the early part of this century, as may be gathered from the following statement of Mr. Alexander Herschel, made in 1867. He remarks that "in England and France, with their dependencies alone, five aërolites in four weeks, about two years ago, were reported in the newspapers to have fallen, and portions of the stones were forwarded to the national museums. Whereas, in 1800, only three fragments of aërolites existed in the Museum of Mineralogy at Vienna, the same collection now contains more than 220 specimens of well-authenticated falls. The gallery of Mineralogy in the British Museum contains a somewhat greater number." In the annual reports of the committee on "Luminous Meteors," presented to the British Association, several newly-observed aërolites, fallen in various parts of the globe, are always recorded. In their volume for 1866, eight decided cases of fallen stones are inserted.

It will be interesting now to give a brief description of a few of the most remarkable aërolites and bolides, selected from a list containing many others of equal scientific and popular interest. Going back at first to the classic authors, we find that one of the most remarkable instances of an ancient fire-ball is that described by Plutarch, in his life of Lysander, as having fallen, in the year 465 B.C., at Ægos Potamos, in the Hellespont, at no great distance from the modern Gallipoli. This stone is also mentioned by the elder Pliny, who saw it about 500 years afterwards. The ancient record says that this aërolite was of a colossal size, that it was as large as a waggon, and that its colour showed that it had been under the action of fire. It is also stated that a large meteor was seen at the time it fell upon the Earth. Diogenes of Apollonia refers to this fire-ball when he remarks that "stars that are invisible, and consequently have no name, move in space together with those that are visible. These invisible stars frequently fall to the Earth and are extinguished, as the stony star which fell burning at Ægos Potamos." There are several recorded falls of aërolites, more or less authentic, scattered over the works of ancient writers, especially in the Chinese annals, which extend backwards to the year 720 B.C., or 255 years before the fall of the aërolite at Ægos Potamos. It is now well understood that the ancient Chinese were great observers of celestial phenomena, and their astronomical observations have been found to be correct, especially their records of the dates of the appearances of comets, meteors, and the recurrence of eclipses. In the *Chun Tsew*, a historical work written by Confucius, there are notices of no less than thirty-six solar eclipses, the first of which was observed on the 22nd of February, 720 B.C., and the last on the 22nd of July, 495 B.C. In another work, entitled *Tung Keen Kang Muh*, in 101 volumes, there are numerous records of eclipses, comets, and falling stars, in the order they occurred from the year 481 B.C. to the epoch of the Christian era.

I I

An authentic fall of a remarkable meteoric stone took place on the 7th of November, 1492, at Ensisheim, in Alsace. In the *Shepherd's Calendar*, a very curious and rare book, printed in 1506, a copy of which is in the British Museum, this fall of an aërolite is thus related:—"Shepardys that lyes the nyghtys in the feldes do se many *Impressions* in the ayer above the erthe, that they that lythe in theyr beddys sees not . . . Lo you people ye may se that these Impressyons be very marvelous, and yet some Ignorante people wyll not beleve it, and wyll thynke it unpossybyll; but you shalle understande that in the yere of oure Lorde a thousand cccclxxx and xii, the vii daye of November, there felle one thynge mooste marvelous in the shyre of ferrat: it happenyd in the dukedome of Autryche, by a towne namyd *Ensychyne*, and on the daye beforsayd felle a grete and orybyll thonder in the feldys, and there felle a greate Thonder Stone, the whiche dyd way ccxl pounde and more, the whiche stone is there present and kept yet in the sayde towne that all maye see it that wyll come: of the whiche Stone here foloweth the eppataffe wreton underneath it." This inscription appears to have been written in Latin, German, and French. It has been rendered into English as follows:—

> "In fourteen hundred and ninety-two
> There happen'd here a great ado,
> For close without, before the town,
> The seventh of November's moon,
> A stone was fall'n, and there it lay,
> With thunder, and in open day!
> Two hundred and a half it weigh'd;
> Its colour iron. Then they made
> Procession, and 't was hither borne;
> But much by force from it was torn."

Soon after the fall of this aërolite at Ensisheim, the Emperor Maximilian I. happened to be present with his army in this part of the continent, when information was conveyed to him that an extraordinary fire-ball had recently descended from heaven. He immediately ordered a detailed account of the phenomenon to be drawn up for his inspection. He then became so much interested in this aërolite that he caused it to be suspended by a chain in the cathedral of Ensisheim, with the above inscription attached to it. Here it remained undisturbed for a period of three centuries, when, during the great French revolution, it was carried off to Colmar, and several pieces were broken from it. One of these fragments may still be seen in the museum of the Jardin des Plantes, Paris, and a small one of about one pound in weight is deposited in the British Museum. The original weight of this extraordinary meteoric mass was 270 pounds, and it is the earliest authentic aërolite of which portions may be seen at the present time.

An immense mass of iron-stone found in Siberia, in 1776, by Dr. Pallas, a travelling naturalist, forms the subject of the well-known paper by Chladni, to which we have already alluded. This curious meteoric mass was found on a mountain near the river Jenesei, and was composed of totally different substances from those of the neighbouring rocks. It seemed impossible for it to be a work of art, and it was unlike any iron-ore seen before or since. It was held in great veneration by the Tartars, among whom a popular tradition prevailed that it had fallen from the stars. In 1779, this block of iron-stone was removed to the town of Krasnojarsk, by the superintendent of the iron mines in that locality. It was of an irregular shape, and its weight 1,400 pounds. Its texture was not solid, but cellular like a sponge, the cells containing small granular particles of a glassy nature, which were subsequently found to be simple mineral olivine. The analysis of the stone, by Howard, showed that it contained 17 per cent. of nickel. Klaproth and John found a less quantity of nickel, while Laugier's experiments showed that the portion analysed by him was composed chiefly of silica, magnesia, sulphur, and chromium. It would, therefore, appear that its composition was not of a uniform character. Other masses of meteoric iron have been found in various parts of the globe: one in the jurisdiction of Santiago del Estero, about 500 miles north-west of Buenos Ayres, consisting of 90 per cent. of iron and 10 per cent. of nickel, specimens of which have been placed in the British Museum. Stones of the same metallic nature have also been found at the Cape of Good Hope, near Bahia, in Brazil, and at Agram, in Croatia. At Agram the meteoric body was actually seen to fall to the Earth, and was considered for many years to be the only iron-stone which had visibly descended from the atmosphere. The sky was quite free from cloud when this iron mass was seen shooting along the heavens from west to east, making a hollow noise as it proceeded. At the explosion of the meteor, a very loud report was heard, accompanied by a great smoke, from which two masses of iron, welded together like a chain, were precipitated with great velocity to the ground. A fragment of this aërolite is also preserved in the British Museum.

A splendid specimen of meteoric iron has been discovered near the Rio Florido, Mexico, at a place called Concepcion. It is four feet above the surface of the ground. Its shape is regular, being from two to three feet in one direction, and varying from two to five in the other. Where worn by the rubbing of hands of passers-by, it is bright, and has the appearance of being nearly pure metal. A steel hatchet cut into it easily, and a neighbouring blacksmith has cut off pieces for horse-shoes. It is supposed to weigh five or six thousand pounds.

It is not always easy to discriminate between the true aërolite fire-ball and an ordinary bolide. There is, however, a sensible difference in the phenomena after the explosion has taken place. For example, in addition to their brilliant and rapid transit across the heavens, aërolitic fire-balls generally burst with a loud report, and at the same time fragments of various sizes are cast down on the Earth in the form of a shower of stones. Bolides, on the contrary, although occasionally bursting with a similar report, are of a much more silent character. They are of a much looser texture than aërolites, and are composed of substances evidently much more inflammable. As a bolide flits from one quarter of the sky to another, it burns with intense brilliancy in all varieties of colour, frequently illuminating the landscape in every direction; but no stones of any magnitude are usually precipitated at the moment of extinction. During the great November star-shower of 1866, several ordinary examples of the bolide class were exhibited. Some bolides have been observed as large as the Moon, making full allowances for the temptation to exaggerate when viewing any unexpected celestial phenomenon.

The great meteor which passed over a portion of England and France near midnight, on October 7, 1868, was one of the most magnificent bolides of recent years. The light emitted from it was so brilliant, that it not only completely overpowered the Moon, but it also cast strong shadows on the ground. It was accompanied by a flaming tail of great length, and was observed over a large extent of country, accounts of its appearance having been received from Keynsham and Clifton, near Bristol, Oxford, various parts of London, Brighton, Ramsgate, Dover, Dunkirk, Havre, Rouen, Paris, and other intermediate places. The meteor was seen as far as Dusseldorf, in Rhenish Prussia. The different descriptions sensibly agree, except as to its direction, from which we may conclude that the whole of the south of England and north of France was, for a few seconds, brilliantly illuminated at the same moment. Most of the accounts mention the quarter of the heavens towards which the meteor was travelling, but little can be gathered from them when some of the observers saw it going east, others west, a few south, while our neighbours, the French, remarked that it went north. We are inclined to believe, however, that, as viewed from London, the meteor had a nearly horizontal motion inclined slightly towards the south-eastern horizon, within a few degrees of which it disappeared. Little confidence can also be placed in the reputed observations of its fall to the ground. One of our correspondents states that he distinctly saw it descend in or near Nunhead Cemetery; another mentions Blackheath, while a third felt certain that it fell in Sussex. Several places in France have also been named as spots where the meteor was seen to fall. We have no doubt whatever, that all these supposed falls are mere fancies in the minds of unpractised observers. It is very probable that in such a brilliant bolide the combustion of the composing matter was so perfect, that little was left at its extinction, excepting very small fragments, which might be precipitated to the Earth unobserved. Unfortunately the meteor does not appear to have been seen by scientific observers, competent to fix its path in the heavens with respect to the stars. Its general appearance can, however, be gathered from the following remarks:—

From Wimbledon the bolide was seen to consist of "a red ball, emitting bright sparks, and having a flaming tail of great length, illuminating the spot where the observer stood with great brilliancy, much as a flash of lightning might do. The colour of the flame was bluish. The sky was perfectly clear at the time, and the Moon was shining brightly; but the light of the meteor, which lasted several seconds, completely overpowered that of the Moon . . . . . . The observer was standing with his face the other way when the sudden light made him turn round, and the latter part of the meteor's flight was seen through trees, but so distinctly as if a firework were behind the trees." Another account states that in St. Paul's Churchyard everything was as clear as day, the cathedral and houses at the north-west corner of Cannon Street standing out in bold relief against a brilliant sky, the lights in the gas-lamps being for the time invisible. An eye-witness at Rouen has described the meteor as "an incandescent globe which, in bursting, split into two tongues of fire, and in this state fell into the Duke of Almazan's pleasure-grounds." Another in Paris saw the bolide pass between the two stars Beta and Gamma Ursæ Minoris, gradually increasing in size till it reached the apparent diameter of the Moon, when it exploded. The report of the detonation was not heard till 5 minutes 28 seconds after the disappearance of the meteor. If this interval of time be correct, it would appear to have burst somewhere between Paris and the English Channel, and at least sixty miles from the observer. At the moment of the explosion, the bolide looked like a brilliant electric light; it then changed suddenly to a bright red, then blue, afterwards yellow, and finally to a green colour before the extinction. The report of the explosion was undoubtedly heard by many other persons in the departments of Seine, Somme, Aisne, and Oise; but in all cases the bolide was at some distance. It has been asserted in a newspaper paragraph that it fell at La Varenne, close to the Vincennes Railway, and that it was found to measure a metre in length. Belleville, near Paris, has been also mentioned as the spot where this meteor fell. Notwithstanding these reputed falls, we believe, as we have previously stated, that nothing reliable is known of the descent of any fragmentary portions of this meteor. As a bolide, however, the remembrance of its magnificence will not soon pass away.

Mr. R. P. Greg, of Manchester, has catalogued all authentic bolides and showers of stones observed, from the commencement of the Christian era. The separate entries amount to 2,400. From this elaborate compilation of Mr. Greg, many interesting falls might be selected as illustrations of these extraordinary fiery meteors, but there is so very much in common observed in the appearance of all large bolides that it is not necessary to give more than a brief notice of a few striking examples of fiery meteors which have occurred during the last and present centuries.

The first is that extraordinary blazing meteor seen all over England, on the evening of March 19th, 1719, and which was supposed to be at least a mile in diameter. A full description of its appearance was drawn up by Dr. Halley. This meteor was observed at London, Oxford, several places in Cornwall and Devonshire, Worcester, Kirkby Stephen on the borders of Westmoreland and Yorkshire, Aberdeen, Paris, Brest, and other places. Dr. Halley computed its height to be about sixty geographical miles. A detonation like thunder shook the houses as it passed. In the south-western counties the explosion was heard as the report of a very great cannon, or rather of a broadside, at a distance, followed by a miscellaneous noise, as if a volley of small arms had been discharged. "What was peculiar to this sound was, that it was attended with an uncommon tremor of the air, and everywhere in those counties very sensibly shook the glass windows and doors in the houses, and, according to some, even the houses themselves, beyond the usual effect of cannon, though near; and Mr. Cruwys, at Tiverton, on this occasion lost a looking-glass, which being loose in its frame, fell out on the shock, and was broken." In London, the light emitted from the meteor was little inferior to that of the Sun, candles gave no light, the stars disappeared, and the Moon, then nine days old, was scarcely visible. The general appearance all around was that of perfect day.

Another meteor of historic interest was observed near Benares, on December 19th, 1798, near 8 P.M. It had the form of a large ball of fire, accompanied by a noise resembling thunder. The noise was followed by a shower of heavy stones. They reached the Earth with a great velocity, penetrating the ground to a depth of six inches. The sky at the time was perfectly clear, as it had been for some days previously. The stones belonging to this aërolite were of different sizes, the most perfect having the form of an irregular cube, rounded at the edges, but the angles were for the most part preserved. They varied from three to more than four inches in length; one of the largest weighed two pounds eleven ounces. The stones were covered with a hard and dark incrustation, resembling in some places a kind of varnish. Most of them were partially fractured, probably from the effects of the concussion on reaching the ground, or by the striking of one stone against another. They also appeared to have been subjected to an intense heat. This is the same meteor as that analysed by Howard, the details of which are given in the *Philosophical Transactions* for 1802.

A magnificent bolide, visible over the whole of England, from Northumberland to the Land's End, was very successfully observed at 10^h 45^m P.M., on February 11th, 1850. In appearance it was very similar to that of October 8th, 1868; but its path among the stars was so carefully noted, that to the scientific observer it exceeds even that celebrated meteor in interest and importance. The accounts from qualified observers received at the Royal Observatory and by Mr. Glaisher were very numerous. One from the Rev. J. Jordan, of Enstone, Oxfordshire, gives a description of the meteor as seen in his neighbourhood. He had been in bed about ten minutes when a vivid flash of light shone upon his window, remaining for several seconds. Shortly afterwards a sound was heard as of a dreadful explosion, which shook the window of the room and the whole of the house. He says:—"The next morning the noise of the explosion, as well as the light which had been seen, were in every one's mouth; several thinking, as I had done, that some persons had come under their windows with a dark lantern. My own servants were greatly alarmed, and believed that a violent entry had been made into the house, mistaking the explosion for the falling of a heavy shutter. Some of my neighbours, on hearing the noise, fancied that barrels had burst in their cellars, others thought that a door or shutter was broken open and thrown down; and indeed the alarm was general, so loud was the noise, and so peculiar the sound as differing from that of thunder." Another observer remarks, that the meteor first appeared about four times as large as Venus, with a dark golden lustre. It then rapidly increased in magnitude, till the whole heavens were lit up with brilliancy. The light exceeded that of the full Moon. At Hartwell, near Aylesbury, it seemed like a great mass of fire darting across the sky from west to east; a report like thunder followed about two or three minutes after the extinction.

This great meteor was observed at Greenwich, Brighton, Bath, Penzance, Birmingham, Rugby, Nottingham, Chester, Hull, Durham, and at several intermediate places. The observations showed that it passed over Shropshire, Warwickshire, Northamptonshire, to the vicinity of Biggleswade, in Bedfordshire, where it exploded. At its first appearance its altitude above the Earth was about eighty-four miles, and at its extinction nineteen miles. The actual diameter of this fiery globe was supposed to be from 1,800 to 2,000 feet, or one-third of a mile. Its tail, or stream of light, attached to the head or nucleus, was of great brilliancy, and from the notes of the observers, must have been several miles in length.

A brilliant detonating meteor observed by Mr. Warren De La Rue near Cranford, Middlesex, on the evening of November 21st, 1865, deserves a special mention. This beautiful meteor rose from the eastern horizon surrounded by a halo of light similar to a comet. When about forty degrees high, it emerged like a fire-ball from a Roman candle, with a blue colour, and about a quarter of the size of the Moon. This meteor traversed the entire length of the valley of the Thames, performing a distance of seventy-five miles from east to west in about six seconds and a half. Mr. Penrose, of Wimbledon, heard a loud report, like that of a cannon fired off at some miles distant, about two minutes and twenty seconds after the meteor had disappeared.

Some of the largest meteors have been observed in broad daylight; that which we have previously mentioned as having fallen at L'Aigle, Normandy, is one of the most remarkable of them. On April 26th, 1803, the sky being clear, a small rectangular cloud was seen to move with great rapidity over various places in Normandy, some of which were far distant from the others. At L'Aigle loud explosions were heard, resembling the sound of cannon and musketry, at the time this cloud-like phenomenon was apparently overhead. Soon after, a loud hissing noise was heard, and stones were seen to fall to the ground. It was subsequently discovered that the stones were scattered over a space two and a half leagues long by one broad. Above two thousand were collected, varying in weight from two drachms to seventeen and a half pounds. From an analysis of one of these stones by Thénard, it was found to contain equal proportions of silica and iron, of 45 per cent. each, with small quantities of magnesia, nickel, and sulphur.

The last meteor which we need describe is also one observed in the daytime. It was evidently one of the largest class of bolides. It appeared about a quarter to eleven on the morning of June 20th, 1866, and was visible as a very brilliant object from various parts of England and the continent, principally, however, in the counties of Kent and Sussex, at Boulogne, Calais, and Lille, in France, and as far as Delft in Holland. From a comparison of several observations, this meteor is supposed to have exploded over some part of the district between the towns of Boulogne and St. Omer. At Boulogne the concussion of the air produced by the explosion was universally felt within the houses. Mr. Francis Galton, then one of the secretaries of the British Association, who was residing at that time at Boulogne, heard the explosion, and noticing that the people in the hotel were running about the house and into the courtyard in a state of alarm, asking one another what had happened, and observing also that crowds were collecting in the street, he went out just in time to see the long, narrow, smoke-like train of a meteor hanging in the sky, the average breadth of the train of smoke being about one degree. At Wrotham, Kent, a shock similar to that of a heavy body falling overhead shook the houses and windows, and startled labourers in the fields. There were two reports at this place, one apparently coming from the S.W., and the other in the opposite direction, which seemed like an echo of the first. At Dover Castle a loud report was heard thirty seconds after the disappearance. Mr. Nasmyth saw the

THE DAYLIGHT METEOR OF JUNE 20TH, 1866.

meteor from his residence at Penshurst distinctly pass across a clear break in the sky, and then disappear behind a mass of clouds. He heard no report of the explosion. Mr. Nasmyth's observations will give the reader a clear conception of this remarkable daylight bolide. He remarks that, "while walking in my garden about a quarter to eleven on the forenoon of June 20th, I was startled to see a bright red comet-shaped object rapidly moving across the clear blue sky about 35° above the horizon. The length of the meteor was about 1°, or twice as long as the Moon appears in diameter. The motion was majestic, yet rapid, for it traversed a space of 80° in rather less than two seconds. The direction was from N.W. to S.E. The advancing end of the meteor was brilliant red, with a white or shining envelope or head; the after part, or tail, was a ragged fan-shape, with a waving motion, accompanied by white vapours, and followed by a faint white vapour-tail. It disappeared from my sight behind a mass of clouds, and I listened for some time to catch any report or sound of explosion, but I heard none. The passage of the meteor was nearly parallel to the horizon, but with a slight dip or decline to the S.E. It is impossible to convey by words the impression left by the appearance of this mysterious object, majestically traversing the clear blue sky during brilliant sunshine. Had it made its appearance at night, the whole of England would have seen more or less of its light."

According to the researches of the celebrated Swedish chemist, Berzelius, the chemical elements of which the greater part of these meteoric masses are composed are very similar to those which are found throughout the Earth's crust, including iron, nickel, manganese, cobalt, chromium, copper, zinc, arsenic, soda, potash, phosphorus, sulphur, and carbon, or about one-third of all the known simple bodies. In some aèrolites additional substances have been found to a more or less extent. The mode in which the constituent parts of aèrolites are compounded gives them, however, the general aspect of something foreign to our telluric rocks and minerals. But this is more apparent than real, for we have stated in a previous chapter that there is every reason to believe, from the spectrum analysis of the light of the heavenly bodies, that not only meteors, but all the stars and other objects moving in space have been formed, more or less, from materials analogous to those which are to be found in our own terrestrial globe.

"When I consider the heavens, the work of Thy hands—the Moon and the stars, which thou hast ordained—Lord, what is man that Thou art mindful of him, or the son of man that thou regardest him?" Feelings similar to those expressed by the Psalmist have probably entered occasionally into the minds of all, more or less, who contemplate the aspect of the heavens, either with the unassisted eye, or by the space-penetrating telescope. If we compare ourselves individually with all this vast and unknown infinitude of space above and around us, it is natural to feel overwhelmed with the magnitude of the wisdom displayed at this unbounded creative power of God, and of our comparative individual insignificance. Dr. Whewell, in his excellent *Bridgewater Treatise*, remarks that "if, endeavouring to trace the plan of the vast labyrinth of laws by which the universe is governed, we are sometimes lost and bewildered, and can scarcely, or not at all, discern the lines by which pain, and sorrow, and vice, fall in with a scheme directed to the strictest right and greatest good, we yet find no room to faint or falter; knowing that these are the darkest and most tangled recesses of our knowledge; that into them science has as yet cast no ray of light, that in them reason has as yet caught sight of no general law by which we may securely hold: while, in those regions where we can see clearly, where science has thrown her strongest illumination upon the scheme of creation; where we have had displayed to us the general laws which give rise to all the multifarious variety of particular facts; we find all full of wisdom, and harmony, and beauty: and all this wise selection of means, this harmonious combination of laws, this beautiful symmetry of relations, directed, with no exception which human investigation has yet discovered, to the preservation, the diffusion, the well-being of those living things which, though of their nature we know so little, we cannot doubt to be the worthiest objects of the Creator's care."

Recent observations have shown us that the evidence is now decisive as to the existence of an almost perfect analogy between the elementary materials composing the various celestial bodies, all separated from each other by distances too vast for our conceptions. If, then, to this fact, we add, with perfect certainty, that the peculiar movements of the binary stars indicate clearly that the same laws of universal gravitation which govern the motions of the Earth and planets around the Sun exist in the remotest regions of the universe, one cannot avoid believing that all this order and regularity must have been produced by some guiding hand, even by the one Great Creator of the heavens and the earth, and all that is contained in them. We have no faith in the opinions of those philosophers who throw doubts on this subject, and who refer the origin of all things to natural causes, which they are, however, unable to explain. As an astronomical observer for more than thirty years, the author may perhaps be permitted to remark, that every year adds fresh conviction to his mind that all this order in the formation and movement of the stars and planets could not possibly have been the result of simple natural causes without the controlling hand of a governing Architect of the universe. In his opinion, humble though it may be, he cannot believe for one moment that the "starry hosts" above us have come to us by mere *chance*; but, on the contrary, that they all exist for some wise purpose, probably unknown to us, through the unerring will and power of Him who "spake, and it was done;" who "commanded, and it stood fast."

> "Oh! who can lift above a careless look,
>   While such bright scenes as these his thoughts engage,
> And doubt, while reading from so fair a book,
>   That God's own finger traced the glowing page;
> Or deem the radiance of yon blue expanse,
>   With all its starry hosts, the careless work of CHANCE!"

ASTRONOMY AND THE BIBLE.

"Of old hast thou laid the foundation of the earth ;
And the heavens are the work of thy hands.
They shall perish, but thou shalt endure ;
Yea, all of them shall wax old like a garment ;
As a vesture shalt thou change them, and they shall be changed :
But thou art the same, and thy years shall have no end.
The children of thy servants shall continue,
And their seed shall be established before thee."

Psalm cii. 25–28 ; *Heb.* i. 10–12.

ASTRONOMY AND THE BIBLE.

"THE heavens declare the glory of God; and the firmament showeth his handy-work." Devout men in all ages have delighted in the contemplation of Jehovah's majesty and power, as displayed in the Sun, Moon, and stars, which he has created. The psalmist David, among other sacred writers, may be quoted for the delight and pious awe with which he dwelt upon the sublime subject of astronomy. "By the word of the Lord were the heavens made; and all the host of them by the breath of his mouth." "Of old hast thou laid the foundation of the earth; and the heavens are the work of thy hands." "Ye are blessed of the Lord which made heaven and earth. The heaven, even the heavens, are the Lord's; but the earth hath he given to the children of men." "When I consider thy heavens, the work of thy fingers, the moon and the stars, which thou hast ordained; what is man, that thou art mindful of him? and the son of man, that thou visitest him?"

The prophet Isaiah, too, stands out prominently among those who gathered arguments for his most solemn appeals, by references to the wisdom and might and goodness of the Lord as exhibited in the firmament above, as well as on the earth beneath. "Thus saith the Lord, thy Redeemer, and he that formed thee from the womb, I am the Lord that maketh all things; that stretcheth forth the heavens alone; that spreadeth abroad the earth by myself." "Thus saith God the Lord, he that created the heavens, and stretched them out; he that spread forth the earth, and that which cometh out of it; he that giveth breath unto the people upon it, and spirit to them that walk therein." "Who hath measured the waters in the hollow of his hand, and meted out heaven with the span, and comprehended the dust of the earth in a measure, and weighed the mountains in scales, and the hills in a balance?" "It is he that sitteth upon the circle of the earth, and the inhabitants thereof are as grasshoppers; that stretcheth out the heavens as a curtain, and spreadeth them out as a tent to dwell in."

In like manner many other of the inspired writers, by raising their eyes and their thoughts to the immensity of space above and around the small globe on which they dwelt, obtained clearer conceptions of the God whom they worshipped and proclaimed. "Thou, even thou, art Lord alone," says one; "thou hast made heaven, the heaven of heavens, with all their host, the earth, and all things that are therein, the seas, and all that is therein, and thou preservest them all; and the host of heaven worshippeth thee."

"By his spirit he hath garnished the heavens," adds another. "He stretcheth out the north over the empty place, and hangeth the earth upon nothing." "Which alone spreadeth out the heavens, and treadeth upon the waves of the sea. Which maketh Arcturus, Orion, and Pleiades, and the chambers of the south."

"Ah Lord God," exclaims another, "behold, thou hast made the heaven and the earth by thy great power and stretched out arm, and there is nothing too hard for thee."

The writers of the New Testament, and the apostles of our Lord, were not silent on this interesting theme. "Through faith we understand that the worlds were framed by the word of God." "Thou, Lord, in the beginning hast laid the foundation of the earth; and the heavens are the works of thine hands." "We also," said Paul to the heathens who would have worshipped him and his fellow-apostle for a miracle wrought by them in the name of their Master—"we also are men of like passions with you, and preach unto you that ye should turn from these vanities unto the living God, which made heaven, and earth, and the sea, and all things that are therein."

One or more passages, selected from many others, may fitly be added, as referring us to the great mystery of godliness, God manifest in the flesh. "God, who at sundry times and in divers manners," says the writer of the epistle to the Hebrews, "spake in time past unto the fathers by the prophets, hath in these last days spoken unto us by his Son, whom he hath appointed heir of all things, by whom also he made the worlds." "All things," declares the apostle John, "were made by him, and without him was not anything made that was made." "For by him" [the Lord Jesus Christ], testifies the apostle Paul, "were all things created, that are in heaven, and that are in earth, visible and invisible, whether they be thrones, or dominions, or principalities, or powers: all things were created by him, and for him: and he is before all things, and by him all things consist."

And what of those vast luminaries which have been presented before our wondering thoughts, and described in the preceding chapters: "shall we say of them," asks a modern writer, "that they were created in vain? Were they called into existence for no other purpose than to throw a tide of useless splendour over the solitudes of immensity? Our Sun is only one of these luminaries, and we know that he has worlds in his train. Why should we strip the rest of this princely attendance? Why may not each of them be the centre of his own system, and give light to his own worlds? It is true that we see them not: but could the eye of man take its flight into those distant regions, it would lose sight of our little world before it reached the outer limits of our system; the greater planets would disappear in their turn, before it had described a small portion of that abyss which separates us from the fixed stars; the Sun would decline into a little spot, and all its splendid retinue of worlds would be lost in the obscurity of distance. Why resist the grand and interesting conclusion? Each of these stars may be the token of a system as vast and glorious as the one which we inhabit. Worlds roll in these distant regions; and those worlds must be the mansions of life and of intelligence. In yon gilded canopy of heaven we see the broad aspect of the universe, where each shining point presents us with a sun, and each sun with a system of worlds, where the Divinity reigns in all the grandeur of his high attributes, where he peoples immensity with his wonders, and travels in the greatness of his strength through the dominions of one vast and unlimited monarchy." *

The same writer, after expatiating on the wonders brought to our knowledge by modern astronomy, goes on to ask, " And what is this world of ours in the immensity of space? and what are they who occupy it? The universe at large would suffer as little, in its splendour and variety, by the destruction of our planet, as the verdure and sublime magnitude of a forest would suffer by the fall of a single leaf. The leaf quivers on the branch which supports it. It lies at the mercy of the slightest accident. A breath of wind tears it from its stem, and it lights on the stream of water which passes underneath. In a moment of time the life, which we know, by the microscope, that it teems with, is extinguished; and, an occurrence so insignificant to the eye of man, and on the scale of his observation, carries in it, to the myriads which people this little leaf, an event as terrible and as decisive as the destruction of a world. Now, on the grand scale of the universe, we, the occupiers of this ball, which performs its little round among the suns and the systems that astronomy has unfolded, may feel the same littleness and insecurity. We differ from the leaf only in this circumstance: that it requires the operation of greater elements to destroy us. But these elements exist. The fire which rages within may lift its devouring energy to the surface of our planet, and transform it into one wide and wasting volcano. The sudden formation of elastic matter in the bowels of the earth, may explode it into fragments. The exhalation of noxious air from below, may impart a virulence to the air around us; it may affect the delicate proportion of its ingredients; and the whole of animated nature may wither and die under the malignity of a tainted atmosphere. A blazing comet may cross this planet in its orbit, and realise all the terrors which superstition has conceived of it. We cannot anticipate with precision the consequences of an event which every astronomer must know to lie within the limits of chance and probability. It may hurry our globe towards the sun, or drag it to the outer regions of the planetary system, or give it a new axis of revolution; and the effect would be to change the place of the ocean, and bring another mighty flood upon our islands and continents. These are changes which might happen in a single moment of time, and against which nothing known in the present system of things provides us with any security.

* Dr. Chalmers' *Discourses on Christian Revelation viewed in connection with Modern Astronomy.*

"Now, it is this littleness and this insecurity which makes the protection of the Almighty so dear to us, and brings with such emphasis to every pious bosom the holy lessons of humility and gratitude. The God who sitteth above, and presides in high authority over all worlds, is mindful of man; and though, at this moment, his energy is felt in the remotest provinces of creation, we may feel the same security in his providence as if we were the objects of his undivided care. It is not for us to bring our minds up to this mysterious agency; but such is the incomprehensible fact, that the same Being whose eye is abroad over the whole universe, gives vegetation to every blade of grass, and motion to every particle of blood which circulates through the veins of the minutest animal; that, though his mind takes into its comprehensive grasp, immensity and all its wonders, each one of us is as much known to him as though we were individually the single objects of his attention: that he marks every thought, feeling, and movement within us; and that, with an exercise of power which we cannot comprehend, the same God who sits in the highest heaven, and reigns over the glories of the firmament, is at the right hand of each of us, to give every breath we draw, and every comfort we enjoy."

Well may we say with the psalmist, in adoring humility, "Such knowledge is too wonderful for me; I cannot attain unto it;" and ask in astonishment, with King Solomon, "But will God indeed dwell on the earth? behold, the heaven, and heaven of heavens, cannot contain thee!"

But we may advance still further in our thoughts. What God is doing for and with each of us, he is doing with all the dwellers on earth, and has been and will be doing with all the countless millions who have peopled or will people it, from the commencement of time to its close. "With a mind unburdened by the vastness of all its other concerns, he can prosecute, without distraction, the government and guardianship of every single son and daughter of our species."

"It is, indeed," adds the writer we have quoted, "a mighty evidence of the strength of his arm, that so many millions of worlds are suspended on it; but it surely makes the high attribute of his power more illustrious when we know that, while expatiating at large among the suns and the systems of astronomy, that power is, at the very same instant, impressing a movement and a direction on all the minuter wheels of that machinery which is working incessantly around us." This is what the Almighty Maker of heaven and earth is doing.

And what yet will He do? What is to be said of those remarkable passages in the New Testament, in which are set forth, in prophetic language, at once sublime and terrific, the final doom of the earth we inhabit? Is it credible that this Earth is to be consumed by fire,—that the Sun and Moon are to be darkened, that the stars of heaven are to fall, that the skies are to be wrapped in flame, and to be rolled up as a scroll,—are these oriental figures or dread realities, which, at no distant day, are to strike terror to the inhabitants of the earth?

"I frankly confess," says another writer,* "that I do not know how to answer these questions, and I do not believe that all the science and philosophy which now exist on earth, can fit an individual one particle for their comprehension or solution. There are those who find in the internal structure of the Earth—its volcanoes, with their rivers of molten lava,—evidences of the way in which these sublime predictions are one day to be accomplished. I dare not," adds the writer, "thus point out to the All-Wise the means to accomplish his purposes. I can only bow, and reverently accept what is written."

"And do you believe," it may be asked, "that the day will ever come when this great globe, with its rock-ribbed mountains, shall melt with fervent heat,—its ocean billows flash into unmeasured volumes of fiery steam, — when flaming fire shall wrap the doomed planet, and devour its very being, and blot it from its kindred family of Worlds?"

Here is the answer :—"I know of no special reason why this Earth should be eternal. Its destruction does not involve the well-being of the universe; and were it blotted from existence, it would but momentarily disturb the equilibrium of the great scheme of worlds, of which it forms an insignificant unit."

And when it shall please God to destroy its present form, by a baptism of fire, He can bring it out of this ordeal,—not one atom of matter lost, but all re-modelled, restored, re-created, a new world filled with beauty and joy and perpetual happiness; where death—the wages of sin—shall never appear ; and where neither tears, nor sobs, nor sorrows, shall dim the beauty of its enchanting abodes.

There is yet one more thought which, in this closing chapter, must be briefly touched upon. It relates to the redemption of our otherwise ruined race by the Lord Jesus Christ. However it may be with other worlds, sin has entered into ours, and death by sin. But is it conceivable, it has been asked, that God should pay so much attention to this one world, and set up such wonderful provisions for its benefit as are announced to us in the Christian revelation?—"that the concerns of this puny ball, which floats its little round among an infinity of larger worlds, should be of such account in the plans of the Eternal; or should have given birth in heaven to so wonderful a movement as the Son of God putting on the form of our degraded species, and sojourning amongst us, sharing in all our infirmities, and crowning the whole scene of humiliation by the disgrace and agonies of a cruel martyrdom?"

All this is wonderful; but it is true. And how can we tell that the scenes of Calvary, and the effects of Christ's death, may not have been the salvation of all other worlds, and of myriads of intelligent creatures by whom they are peopled? And if not, "what a grandeur does it throw over every step in the redemption of a fallen world, to think of its being done by Him who unrobed himself of the glories of so wide a monarchy, and came to this humblest of its provinces in the disguise of a servant, taking the form of our degraded species, and letting himself down to sorrows,

* Dr. Mitchel, on *The Astronomy of the Bible.*

and to sufferings, and to death for us! In this love of an expiring Saviour to those for whom in agony he poured out his soul, there is a height, and a depth, and a length, and a breadth" which we cannot comprehend indeed, but which we are called upon to admire and embrace with all the ardour of true repentance, faith, and Christ-enkindled love. Let us never, then, neglect so great salvation, nor lose hold of an atonement made sure by Him who cried that it was finished, and brought in an everlasting righteousness.

"It was not the visit of an empty parade that he made to us. It was for the accomplishment of some substantial purpose, and if that purpose is announced and stated to consist in his dying, the just for the unjust, that he might bring us unto God, let us never doubt of our acceptance in that way of communication with our Father in heaven, which he has opened and made known to us. In taking to that way, let us follow his every direction with that humility which a sense of all his wonderful condescension is fitted to inspire. Let us give ourselves up to his guidance with the docility of children, overpowered by a kindness that we never merited, and a love that is unequalled by all the perverseness and ingratitude of our stubborn nature. For, what shall we render to him for such mysterious benefits—to him who has thus been mindful of us, to him who thus has deigned to visit us?"

May God, by his Holy Spirit, give us grace so to read the book of Nature opened to us in the wonders of astronomy, and making known to us the power of God in creation and preservation, that we may more than ever prize the book of his Love. Here is marvellous mercy! "God so loved the world, that he gave his only begotten Son, that whosoever believeth in him should not perish, but have everlasting life." Be it ours to respond, in devout admiration and gratitude, "I am not ashamed of the gospel of Christ: for it is THE POWER OF GOD UNTO SALVATION."

INDEX.

THE END.